Ceramic Engineering

Ceramic Engineering

Jennifer Carter

www.statesacademicpress.com

Published by States Academic Press,
109 South 5th Street,
Brooklyn, NY 11249, USA

ISBN: 978-1-63989-100-9

Cataloging-in-Publication Data

Ceramic engineering / Jennifer Carter.
    p. cm.
Includes bibliographical references and index.
ISBN 978-1-63989-100-9
1. Ceramic engineering. 2. Ceramic materials. 3. Engineering. I. Carter, Jennifer.
TP810.5 .C47 2022
666--dc23

For information on all States Academic Press publications
visit our website at www.statesacademicpress.com

Contents

Permissions

Index

Preface

Ceramic engineering is a branch of engineering which deals with the creation of objects from non-metallic and inorganic materials, called ceramics. It finds its application in various industries such as aerospace, electronics, automotive, etc. Earthenware, bricks and porcelain are some common examples of ceramics. They have crystalline, partly crystalline or amorphous structure with either short range or long range atomic order. Primarily ceramics are formed by application of heat on raw materials such as clay or any other earthen elements. The general sequence of processes followed in the formation of ceramics is milling, batching, mixing, forming, drying, firing and assembly. Ceramic engineering is an upcoming field of science that has undergone rapid development over the past few decades. This book unfolds the innovative aspects of this field which will be crucial for the holistic understanding of the subject matter. It is a complete source of knowledge on the present status of this field.

To facilitate a deeper understanding of the contents of this book a short introduction of every chapter is written below:

Chapter 1- Ceramic refers to any heat and corrosion resistant material that is hard and brittle, and is made by shaping and then firing a nonmetallic mineral at a high temperature. Some of the topics studied in relation of ceramics are structure of ceramics and particle size distribution. This is an introductory chapter which will briefly introduce all these significant aspects of ceramics.

Chapter 2- Some of the important properties of ceramics are electrical and magnetic properties. The electrical properties that are characteristic for ceramic materials are insulating properties, electrical conductivity, dielectric strength, dielectric constant, etc. This chapter has been carefully written to provide an easy understanding of these properties of ceramics.

Chapter 3- The polycrystalline materials which are produced through controlled crystallization of base glass are referred to as glass ceramics. A few of the topics studied under glass ceramics are kinetic theory of glass formation, structural theories of glass formation and applications of glass ceramics. This chapter closely examines these key concepts of glass ceramics to provide an extensive understanding of the subject.

Chapter 4- Magnetic ceramics are the oxide materials made up of ferrites, which are composed of iron oxide in combination with some other metal. They are crystalline minerals that exhibit ferrimagnetism. It is a type of permanent magnetization. This chapter discusses in detail the different concepts related to magnetic ceramics such as magnetic moment, magnetization, etc.

Chapter 5- There are different types of ceramic materials such as zirconium dioxide, silicon nitride, nanoceramic, coade stone, ceramic matrix composite and transparent ceramics. Zirconium dioxide is a white crystalline oxide of zirconium. All these diverse aspects of these types of ceramics have been carefully analyzed in this chapter.

Chapter 6- Ceramic forming techniques include the various ways used in the formation of ceramics. Some of the techniques of ceramic forming are freeze-casting, freeze gelation and sintering. Freeze-casting is a process that produces materials with complex and three-dimensional pore structures. The topics elaborated in this chapter will help in gaining a better perspective about these ceramic forming techniques.

I owe the completion of this book to the never-ending support of my family, who supported me throughout the project.

Jennifer Carter

Basics of Ceramics

Ceramic refers to any heat and corrosion resistant material that is hard and brittle, and is made by shaping and then firing a nonmetallic mineral at a high temperature. Some of the topics studied in relation of ceramics are structure of ceramics and particle size distribution. This is an introductory chapter which will briefly introduce all these significant aspects of ceramics.

A ceramic is an inorganic non-metallic solid made up of either metal or non-metal compounds that have been shaped and then hardened by heating to high temperatures. In general, they are hard, corrosion-resistant and brittle.

'Ceramic' comes from the Greek word meaning 'pottery'. The clay-based domestic wares, art objects and building products are familiar to us all, but pottery is just one part of the ceramic world. Nowadays the term 'ceramic' has a more expansive meaning and includes materials like glass, advanced ceramics and some cement systems as well.

Traditional Ceramics – Pottery

Pottery is one of the oldest human technologies. Fragments of clay pottery found recently in Hunan Province in China have been carbon dated to 17,500–18,300 years old.

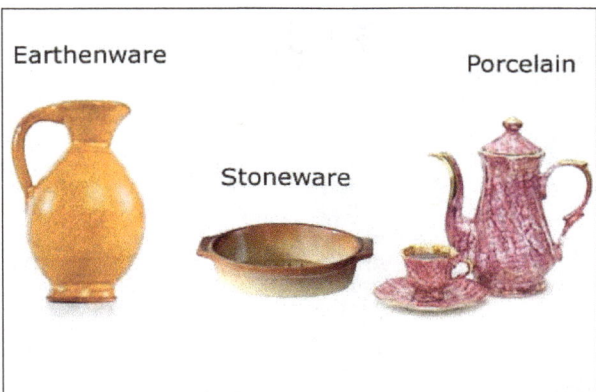

Traditional ceramics are clay–based. The categories of pottery shown here are earthenware, stoneware and porcelain. The composition of the clays used, type of additives and firing temperatures determine the nature of the end product. The major types of pottery are described as earthenware, stoneware and porcelain:

- Earthenware is used extensively for pottery tableware and decorative objects. It is one of the oldest materials used in pottery. The clay is fired at relatively low temperatures (1,000–1,150 °C), producing a slightly porous, coarse product. To overcome its porosity, the fired object is covered with finely ground glass powder suspended in water (glaze) and is then fired a second time. Faience, Delft and majolica are examples of earthenware.

- Stoneware clay is fired at a high temperature (about 1,200 °C) until made glass-like (vitrified). Because stoneware is non-porous, glaze is applied only for decoration. It is a sturdy, chip-resistant and durable material suitable for use in the kitchen for cooking, baking, storing liquids and as serving dishes.

- Porcelain is a very hard, translucent white ceramic. The earliest forms of porcelain originated in China around 1600BC, and by 600AD, Chinese porcelain was a prized commodity with Arabian traders. Because porcelain was associated with China and often used to make plates, cups, vases and other works of fine art, it often goes by the name of 'fine china'.

- To make porcelain, small amounts of glass, granite and feldspar minerals are ground up with fine white kaolin clay. Water is then added to the resulting fine white powder so that it can be kneaded and worked into shape. This is fired in a kiln to between 1,200–1,450 °C. Decorative glazes are then applied followed by further firing.

- Bone china: Which is easier to make, harder to chip and stronger than porcelain – is made by adding ash from cattle bones to clay, feldspar minerals and fine silica sand.

Advanced Ceramics – New Materials

Advanced ceramics are not generally clay-based. Instead, they are either based on oxides or non-oxides or combinations of the two:

- Typical oxides used are alumina (Al_2O_3) and zirconia (ZrO_2).

- Non-oxides are often carbides, borides, nitrides and silicides, for example, boron carbide (B_4C), silicon carbide (SiC) and molybdenum disilicide ($MoSi_2$).

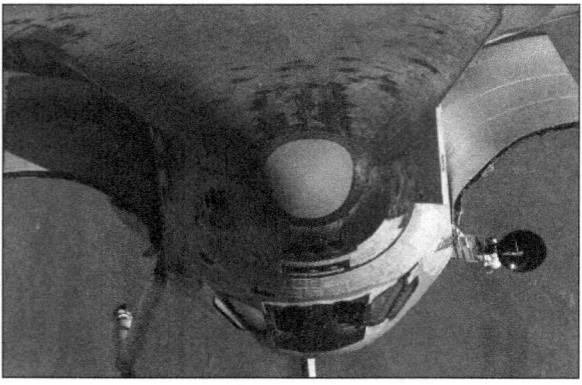

The Space Shuttle Discovery

Part of the space shuttle's outer skin is made up of over 27,000 ceramic tiles. The tiles are designed to withstand the tremendous heat generated on re-entry into the Earth's atmosphere. Production processes firstly involve thoroughly blending the very fine constituent material powders. After shaping them into a green body, this is high-temperature fired (1,600–1,800 °C). This step is often carried out in an oxygen-free atmosphere.

The high temperature allows the tiny grains of the individual ceramic components to fuse together, forming a hard, tough, durable and corrosion-resistant product. This process is called sintering.

Structure of Ceramics

The concepts of lattice, unit-cell and crystal structures will be useful to understand the crystal structures of common ceramic compounds.

Basic Crystallography

Point Lattice

In a point lattice, following characteristics are obeyed:

- There is periodic arrangement of points in space.

- Each point must have identical neighbourhood.

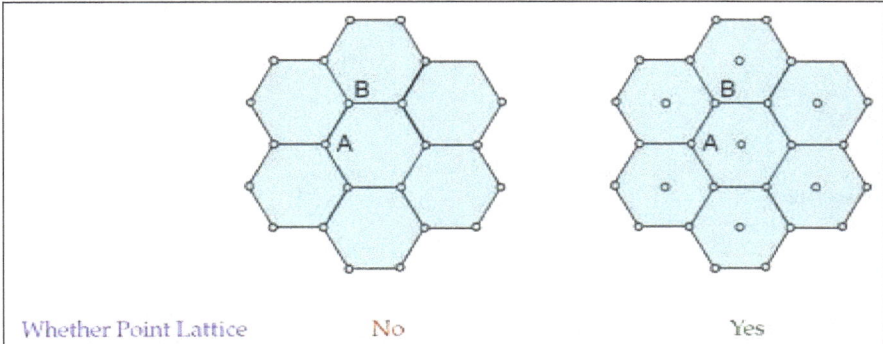

Schematics of arrangement of points in space.

Unit Cell

- A unit cell is the smallest repeatable unit in a point lattice.

- Choice of unit-cell shape is not unique.

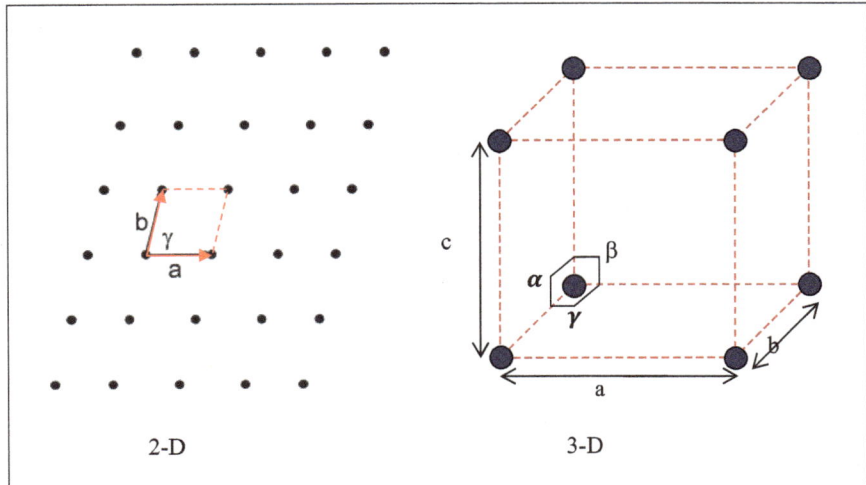

Unit-cell representation.

- Unit-cell parameters for a 3-D unit cells:
  - Axis lengths: a, b and c.
  - Angles: α, β and γ.

Motif and Crystal Structure

- Crystal structure: A Combination of motif and point lattice.

- Motif is defined as a unit or a pattern. For a crystal, it can be an atom, an ion or a group of atoms or ions or a formula unit or formula units. Often, it is also called as Basis.

- When motif replaces points in a periodic point lattice, it gives rise to what is called as a crystal with a defined structure.

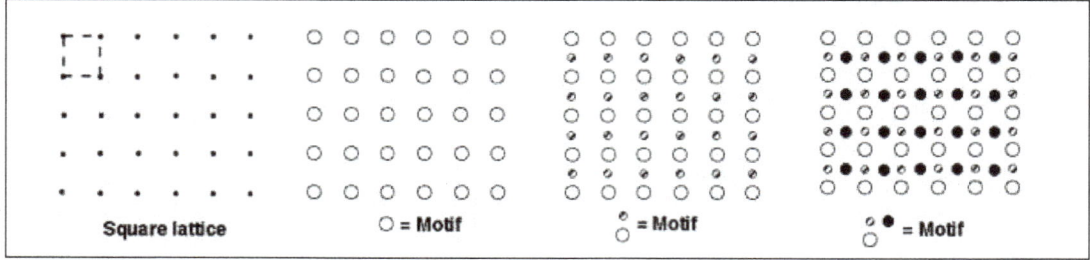

Formation of a periodic crystal structure.

Types of Lattice

Lattice can further be classified into two types:

- Primitive lattice having one formula unit or one lattice point or one unit of motif per unit cell.

- Non-primitive lattices having more than one lattice points or more than one unit of motif per unit cell.

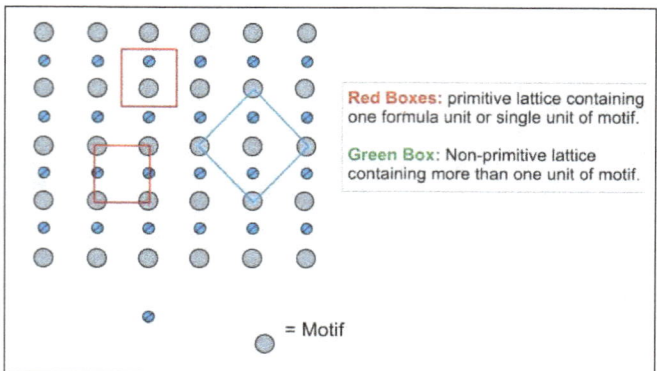

Primitive and Non-primitive lattices.

Crystal Systems

- As you can see that since choice of unit cell is not unique, we can define any unit cell of any shape as long it contains one lattice point.

- However, as one starts defining various shapes, we come up with seven categories, called as crystal systems, in which all possible unit cells shapes would fit provided space filling criteria is fulfilled.

- Seven crystal system are shown below:

Table: Seven crystal systems, the lattice parameters and symmetry requirements.

		Crystal system and lattice parameters	Minimum symmetry
1	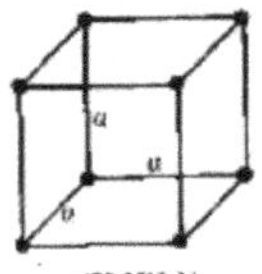 SIMPLE CUBIC (P)	Cubic; a = b = c, $\alpha = \beta = \gamma = 90°$	Four 3-fold rotation.
2	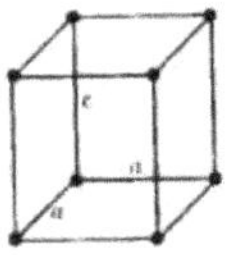 SIMPLE TETRAGONAL (P)	Tetragonal; $a = b^1 c$	One 4-fold rotation (or rotationinversion) axis.
3	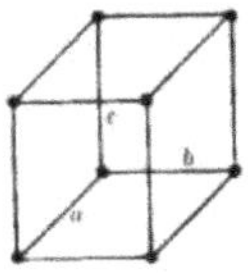 SIMPLE ORTHORHOMBIC (P)	Orthorhombic; $a^1 b^1 c$	Three perpendicular 2-fold rotation (or rotation-inversion) axis.
4	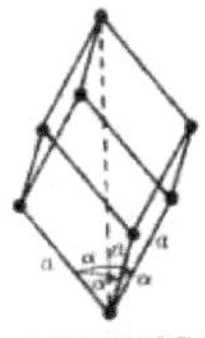 RHOMBOHEDRAL (R)	Rhombohedral; $a = b = c, \alpha = \beta = \gamma^1 90°$	One 3-fold rotation (or rotation-inversion) axis.
5	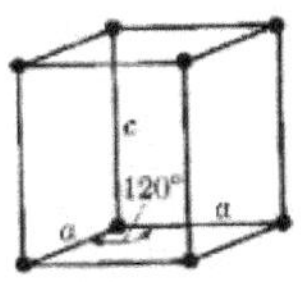 HEXAGONAL (P)	Hexagonal; $a = b^1 c$ $\alpha = \beta = 90° \gamma = 120°$	One 6-fold rotation (or rotation-inversion) axis.
6	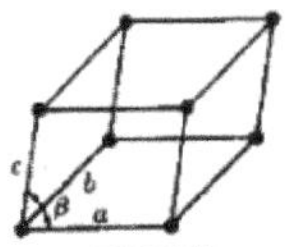 SIMPLE MONOCLINIC (P)	Monoclinic; $a^1 b^1 c$ $\alpha = \gamma = 90°{}^1 \beta$	One 2-fold rotation (or rotation-inversion) axis.

7	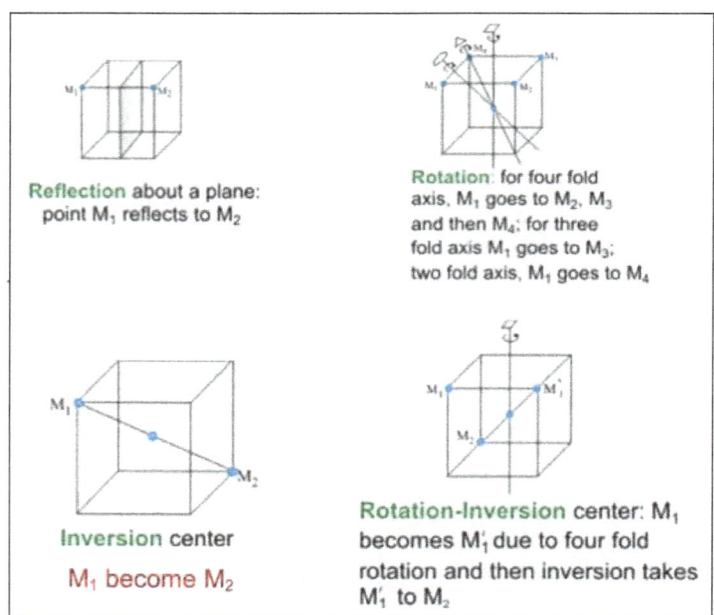TRICLINIC (P)	Triclinic a^1b^1c $\alpha^1\gamma^1\beta^1 90°$	None

Symmetry in Crystals

- The division of crystal structures into various crystal systems and Bravais lattices is based on the symmetry considerations.

- Symmetry is an operation which brings the object back to its original confiscation.

- Symmetry elements underlying a point lattice:

  ○ Reflection: Reflection across a mirror plane.

  ○ Rotation: Rotation around a crystallographic axis by an angle,θ, such as 360°/θ is an integer of value 1, 2, 3, 4 and 6 and is referred as n-fold rotation.

  ○ Inversion: A point at x, y, z becomes its equivalent at (−x,−y,−z)

  ○ Rotation-Inversion: Rotation followed by inversion OR Rotation-Reflection: Rotation followed by reflection.

Reflection about a plane: point M₁ reflects to M₂

Rotation: for four fold axis, M₁ goes to M₂, M₃ and then M₄; for three fold axis M₁ goes to M₃; two fold axis, M₁ goes to M₄

Inversion center
M₁ become M₂

Rotation-Inversion center: M₁ becomes M'₁ due to four fold rotation and then inversion takes M'₁ to M₂

Basic symmetry operations in crystals.

Bravais Lattices

- Taking seven crystal systems and symmetry elements into account, Bravais came out with the fact that there are a total of 14 Bravais Lattices which are shown below.

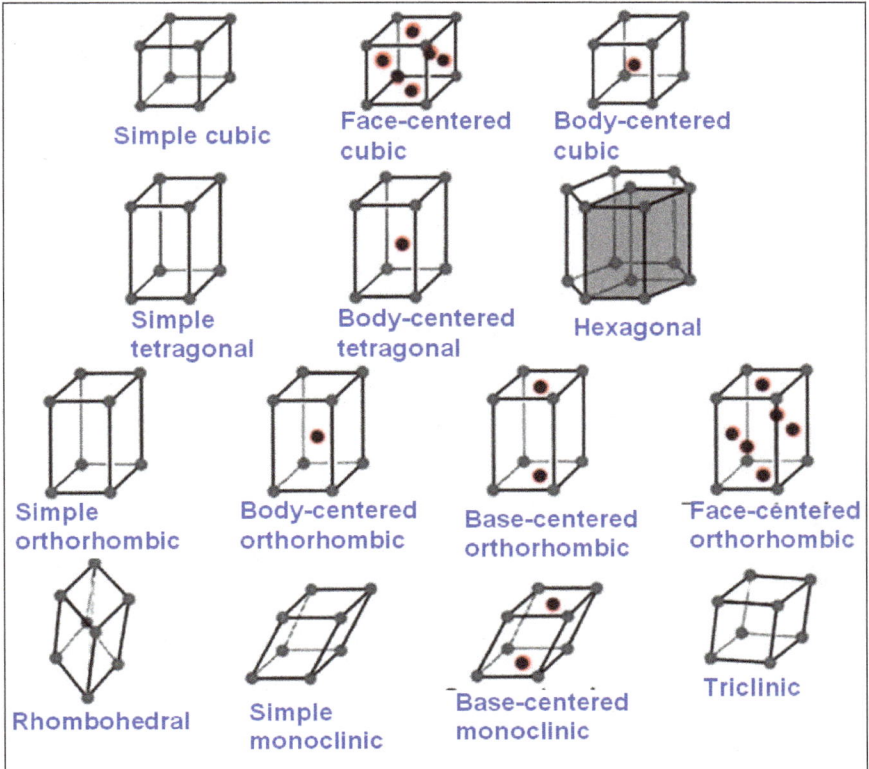

Fourteen Bravais lattices.

Bonding in Materials

Bonding in materials is a very important criterion and determines many of the physical properties of the materials. When two atoms are brought together, the force between the two varies as a function of their separation, r, and can be plotted as shown below in figure. When the separation is large, there is a net attractive force between the atoms, dominated by F_a (negative), and when the separation is small, the atoms repel each other, dominated by F_r (positive). When the atoms are closer, the repulsive forces are high due to repulsion between the outer shell electrons of two atoms and when atoms are taken further away, they drop very rapidly. At a distance r_o, these forces cancel each other and this is called as equilibrium separation between the atoms. At this distance, the potential energy of the system becomes minimum as shown by the lower figure in Figure. The overall potential energy of such a system is represented by:

$$W = \frac{A'z_1z_2e^2}{r} + \frac{B}{r^m}$$

Where left hand side represents the attractive force and right hand side is due to repulsive force. The energy corresponding to the equilibrium separation is called as Bond Energy (W_o) of the materials, typically expressed in kJ/mole or eV/bond. It is the magnitude of this energy which is quite informative vis-à-vis materials properties and the nature of bonding between them. In general, higher the bond energy is, higher would be the melting point, elastic modulus and hardness and lesser is the coefficient of thermal expansion.

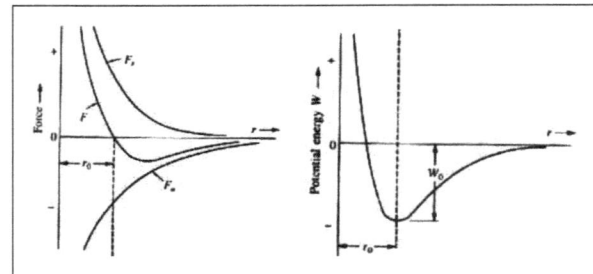

Interatomic forces and potential energy between two atoms in a material as a function of their separation distance, r.

According the magnitude of bond energy, the bonding in materials can be divided in two categories:

- Primary bonding,

- Secondary bonding.

Primary bonds exhibit bond energies in the range ca. 50-1000 kJ/mol. Primary bonds are metallic, covalent and ionic bonds with metallic bond typically being the weakest. On the other hand, bonds with energies lower than 50 kJ/mol are called as secondary bonds. Examples of secondary bonds are van der Walls bond and the hydrogen bond.

Primary Bonding

There are three types of primary bonding mechanisms metallic, covalent and ionic.

Metallic Bonding

This kind of bonding is characterized by presence of a sea of electrons around atoms in metals, also called as free electron gas. The sharing of electrons is not complete enough to provide a covalent character and is delocalized. This very nature allows free movement of electrons around metal cores whilst holding the cores together. This nature gives rise to flexible bonds, good malleability, high electrical and thermal conductivity. Typically most elements to left of fourth column in periodic table show such behavior. Examples of good metals are Ni, Fe, Cu, Au, Ag etc. and their alloys. Most metals show a bond energy below 100 kJ/mole which is a moderate energy. This is why most metals have moderate melting points, moderate elastic modulus and moderate coefficient of thermal expansion.

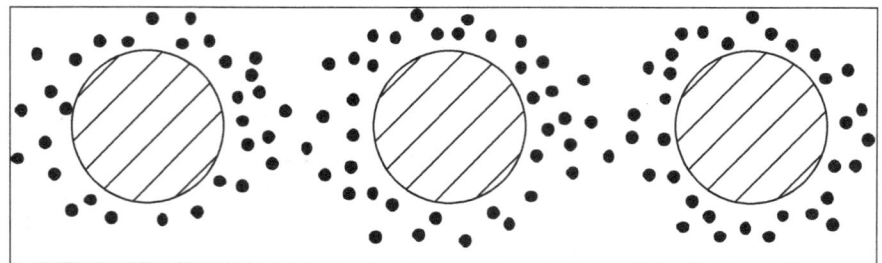

Schematic representation of metallic bonding showing sea of electrons around the metal cores.

Covalent Bonding

In this bonding, atoms share their outer shell unpaired electrons leading to a stronger and directional bonding as shown below. An overlapping of orbitals is required for lowering of potential energy which is typically facilitated by vacant states in the outermost orbitals of bonding atoms. Typically overlapping orbitals are directionally oriented which further gets compounded by hybridization between overlapping orbitals imparting a strongly director character to covalent bonds. For example, 2s and 2p orbitals in Carbon (diamond) form sp hybridization which can hold a total of eight electrons and facilitate sharing between neighboring Carbon atoms in such a fashion that each Carbon atom is surrounded by four Carbon atoms in a tetrahedral coordination forming a regular tetrahedra. This leads to a interbond angle of 109.5° and a strong directionality in diamond. Examples of materials showing this bonding are primarily group IV and V elements and compounds such as Si, C, Ge, and SiC, N, P, As, Sb, and Bi. The bond energies of some of the common covalently bonded materials are 176 kJ/mol for Si and 347 kJ/mol for diamond.

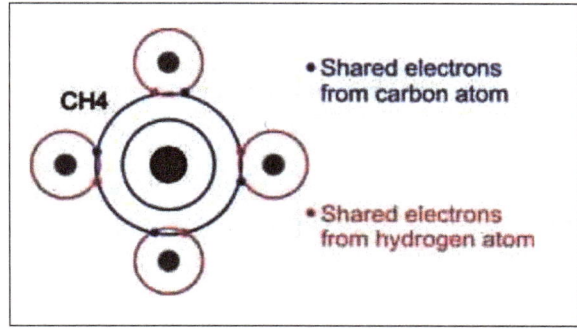

Schematic of Covalent bonding showing sharing of orbitals.

Ionic Bonding

This bonding occurs two oppositely charged elements, for example in NaCl, MgO, ZnO, LiF and many other ceramics and glasses. The electrons transfer from one ion takes place to the other ion. Higher the difference between the tendency to accept and give away the electron(s), described by the difference in the electronegativity of the two ions, stronger the bond will be. Due to spherical symmetry of atoms and thus the bonding force around the atoms, ionic bonding is nor-directional in nature. To begin with ions can be treated as point charges and when the two charges are brought closer to each other, the attractive force between the two can be written by Coulomb's law, i.e., $-Aq_1q_2/r_2$ where A is a constant, q_1 and q_2 are the charges on each ion and r is the separation distance. The resulting potential energy can be written as $-Aq_1q_2/r$. This attractive force varies as square of the distance between the two ions. However, when atoms are brought too close to each other, their electron clouds start overlapping. Since, Pauli's exclusion principle cannot be violated as no two electrons can occupy the same quantum state, this leads to development of a repulsive force between the ions, which increases very sharply with the separation distance. The repulsive energy is expressed as B/r^m where B and m (between 9 to 15) are constant determined empirically. The total potential energy of the system can now be expressed as:

$$W = \frac{-Aq_1q_2}{r} + \frac{B}{r^m} + \Delta E$$

Where ΔE is the difference between ionization potential of cations and electron affinity of the anions and is typically a very small number. The bond energy is determined by taking $\partial W/\partial r=0$ yielding a equilibrium bond energy W_0 at a equilibrium distance of r_0. As the above expressions shows and also true in general, the higher the valence of the ions, the bonding is typically stronger.

This type of bonding typically leads to high bond energies and as a consequence ionically bonded materials typically exhibit high bond strength, high melting point, high elastic modulus, brittle nature and generally low thermal and electrical conductivities making them excellent insulators. Typically bond energies in these materials exceed 100 kJ/mol.

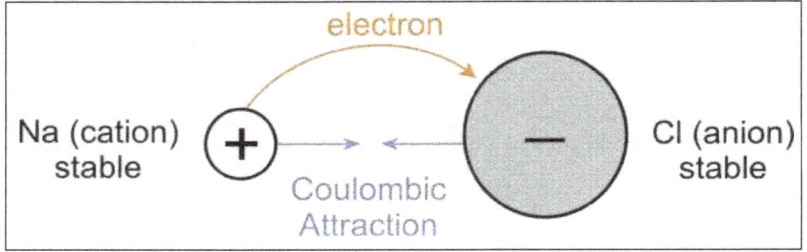

Schematic of ionic bonding.

Secondary Bonding

It arises from the interaction between charge dipoles. The magnitude of this bonding is typically below 50 kJ/mol and even lower. First kind is of between fluctuating dipoles and it is observed in gases like hydrogen as shown below.

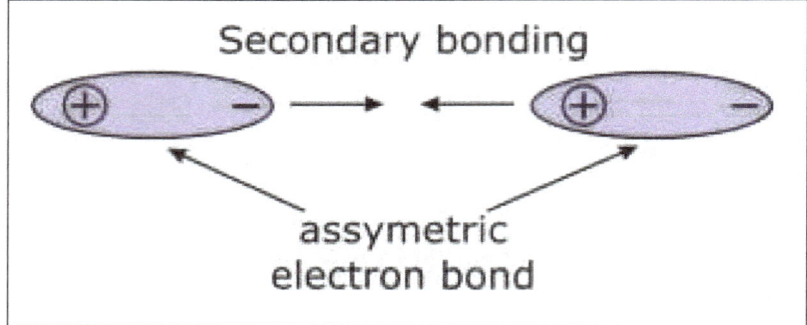

Secondary bonding due to fluctuating dipoles.

Second type is induced by the existence of permanent dipole moment induced due to presence of permanent dipoles in the materials such as on polymers where within a chain units are covalently bonded while between two chains, secondary bonding exists.

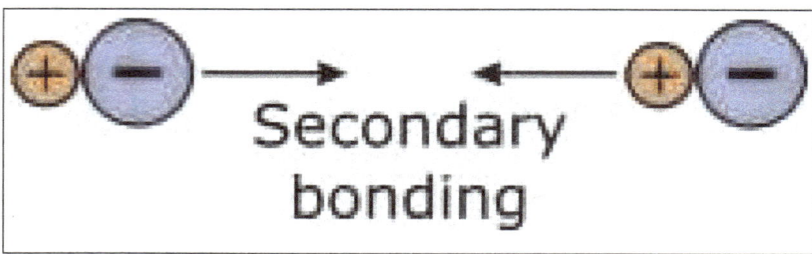

Secondary bonding due to permanent dipoles.

Typical examples of materials (and their bond energies) showing secondary bonding are water (20.5 kJ/mol), ammonia (7.8 kJ/mol) and HF (31.5 kJ/mol).

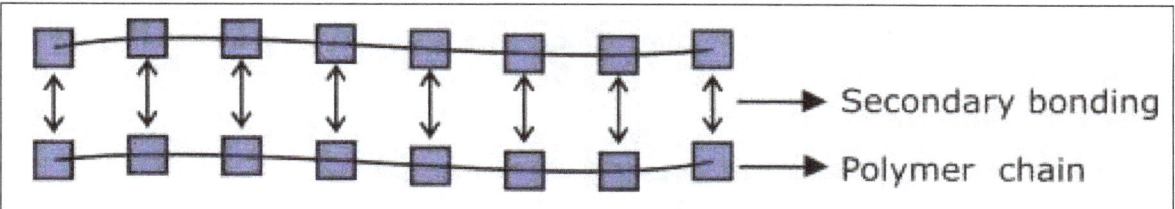

Secondary bonding in polymers.

General Behavior of Materials

In reality most materials show mixed bonding. Nevertheless, the materials are defined by the dominant typed of bonding in them. This classification is useful in approximately classifying their properties and behavior under a given set of conditions. Also, the length of a bond is defined as the distance between centres of two atoms. The materials with stronger bonds typically have smaller bond length, typically below 2 Å, due to high attractive forces. In contrast, secondary bond lengths are typically longer, of the order of 2-5 Å. Although half of the bond length is often taken as diameter of an atom if the two atoms are same, it is often determined by the constraints of packing and coordination number.

Crystallographic Planes and Directions

Faces and directions joining atoms in crystals can be best described by Miller Indices (in the names of W. H. Miller) ascribed to various determine planes and directions. While planes are determined little empirically, directions are nothing but vectors.

Crystallographic Planes

- Identification of various faces seen on the crystal.

- (h, k, l) for a plane or {h, k, l} for identical set of planes where h, k, l are integers.

- A crystallographic plane in a crystal satisfies following equation:

$$\frac{h}{a}x + \frac{k}{b}x + \frac{l}{c}z = 1$$

- h/a, k/b, and c/l are the intercepts of the plane on x, y, and z axes.

- a, b, c are the unit cell lengths.

- h, k, l are the integers called as Miller indices and the plane is represented as (h, k, l).

- Any negative indices in Miller indices of a plane is written with a bar on top such as $\bar{h}$.

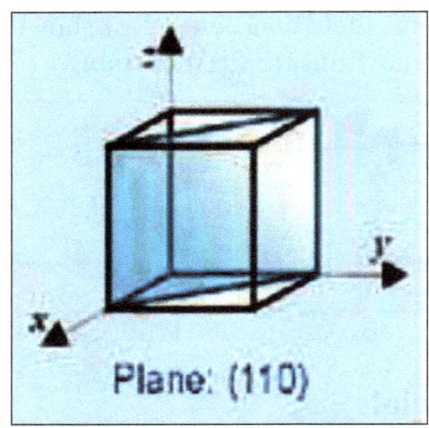

Plane: (110)

How to find Miller indices:

- Find the intercepts of the plane with the crystal axes. Express them as integral multiples of the basis vectors.

- Take the reciprocals of the three integers found in step 1. If possible reduce these to smallest set of integars h, k and l.

- Label the plane (hkl).

Interplanar angle is given by (cubic only),

$$COS\theta = \frac{h_1 h_2 + k_1 k_2 + l_1 l_2}{\sqrt{h_1^2 + k_1^2 + l_1^2} \sqrt{h_2^2 + k_2^2 + l_2^2}}$$

Interplaner spacing is given by (Cubic),

$$d_{hkl} = \frac{a}{\sqrt{h^2 + k^2 - l^2}}$$

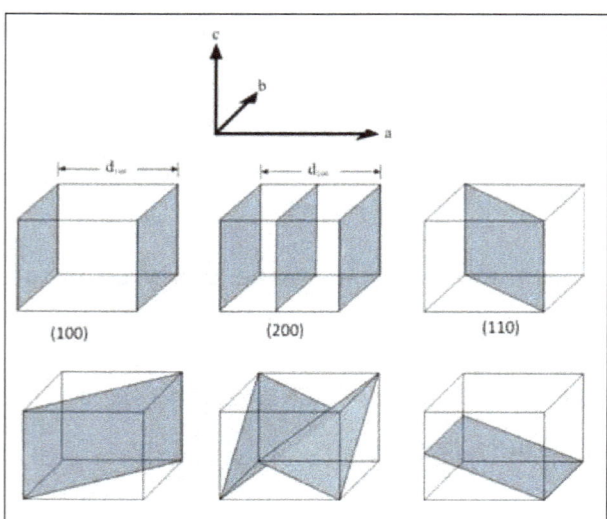

Planes and Directions in crystals.

Directions in the Crystal

- These are basically atomic directions in the crystal.

- Miller indices are [u, v, w] for a direction or <u, v, w> for identical set of directions where u, v, w are integers.

- Vector components of the direction resolved along each of the crystal axis reduced to smallest set of integers.

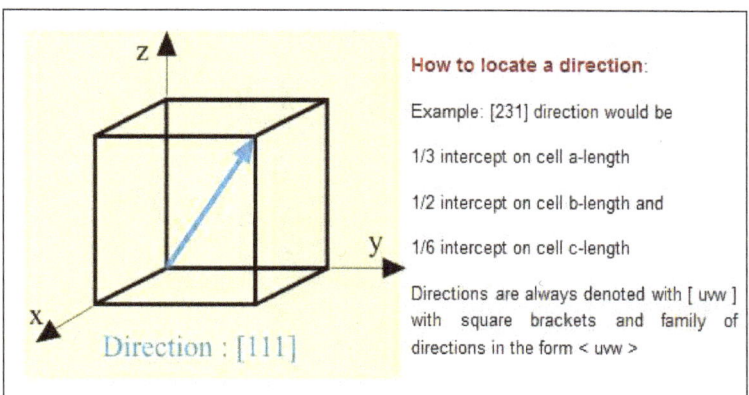

Planes and Directions in crystals.

Packing of Atoms in Metals

- In solids, we consider atoms as hard incompressible spheres which can be packed in various forms in solids. First we will see how atoms pack in metals.

- Atoms in many metals form closed packed structures either in the form of hexagonal closed packed structure or face-centered cubic structures. Some metals are a little loosely packed in the form of body-centered cubic structure. Very rarely atoms pack in metals in the form of simple cubic structure.

Simple Cubic Structure

- Simplest structure crystallographically but in the whole of periodic table, only polonium (Po) possesses this structure.

- Structure contains only one atom per unit-cell.

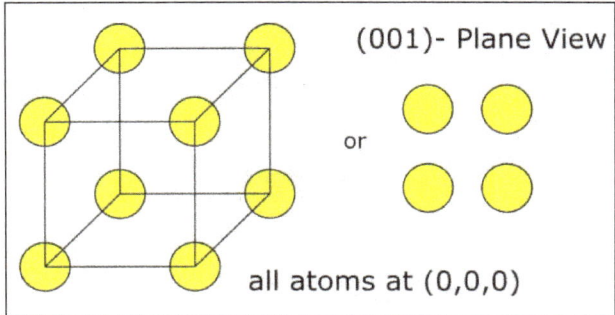

Body Centered Cubic or BCC Structure

- Many metals like W, Fe (room temperature form) possess BCC structure.

- Contains 2 atoms per unit-cell.

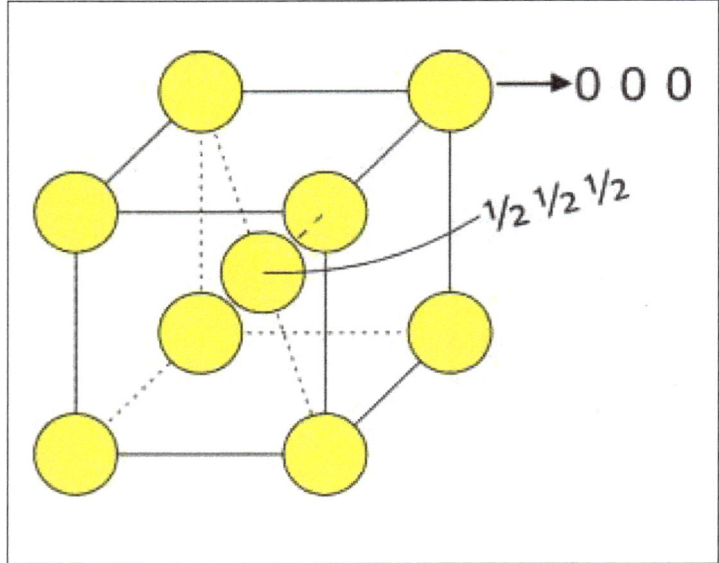

One of the important parameter of interest is packing factor, determining how loosly or densely a structure is packed by atoms.

$$\text{Packing Factor} = \frac{\text{Volume of all atoms in one unit-cell}}{\text{volume of one unit-cell}}$$

If r is the atomic radii in these structures, then

$$\text{Packing Factor (Simple Cubic)} = \frac{1 \times \dfrac{4}{3} \times \pi r^3}{(2r)^3} = 0.52$$

$$\text{Packing Factor (BCC)} = \frac{2 \times \dfrac{4}{3} \pi r^3}{\left(\dfrac{4}{\sqrt{3}} r\right)^3} = 0.68$$

Closed Packed Structures

- Each atoms has 12 nearest neighbours.

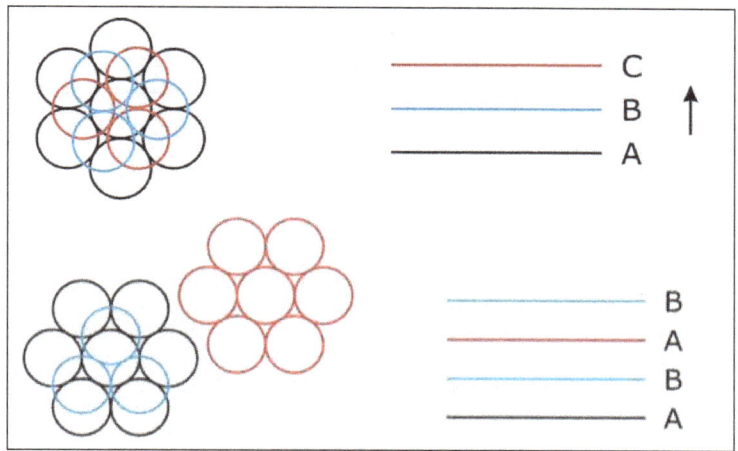

Closed packing of atoms in FCC/HCP metals.

ABC ABC ABC . . . Stacking ⇒ Cubic Closed Packed (CCP) or Face Centered Cubic (FCC) Structure.

AB AB AB . . . Stacking ⇒ Hexagonal Closed Packed (HCP) Structure.

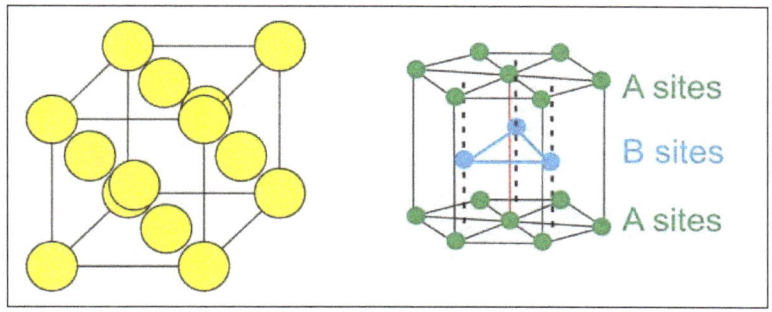

FCC and HCP Structures.

Now you can work out yourself that packing factor of both FCC and HCP is 0.74.

Interstices in Structures

- Since the unit cell is not completely packed as packing efficiency in above structures is less than 100%, there are empty spaces inside which are called as interstices.

- These interstices are very useful because there can contain smaller atoms which modify the properties of materials tremendously, such as carbon (C) in iron (Fe) makes steel and makes iron stronger.

Interstices in FCC Structure

- Tetrahedral Interstices:

  ◦ 2 per atom.

- Octahedral Interstices:

  ◦ 1 per atom.

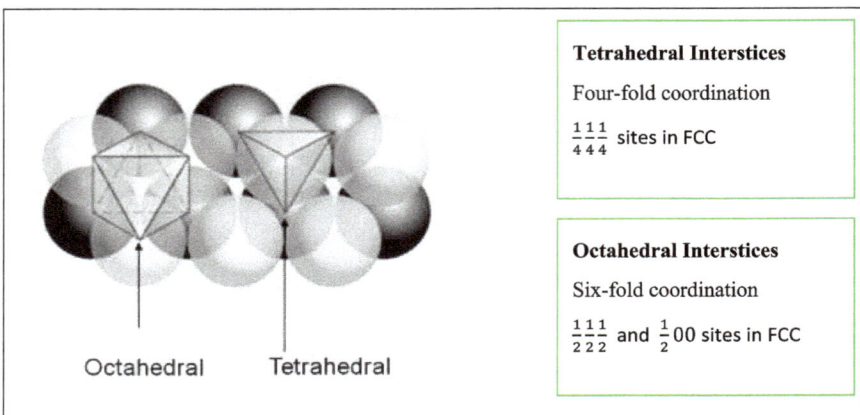

Tetrahedral Interstices

Four-fold coordination

$\frac{1}{4}\frac{1}{4}\frac{1}{4}$ sites in FCC

Octahedral Interstices

Six-fold coordination

$\frac{1}{2}\frac{1}{2}\frac{1}{2}$ and $\frac{1}{2}00$ sites in FCC

Octahedral   Tetrahedral

By simple geometry, you can also estimate the size of the largest interstitial atom that would in these interstices without distorting them,

$$r_{tet} = 0.225 * r$$

$$r_{oct} = 0.414 * r$$

Interstices in BCC Structures

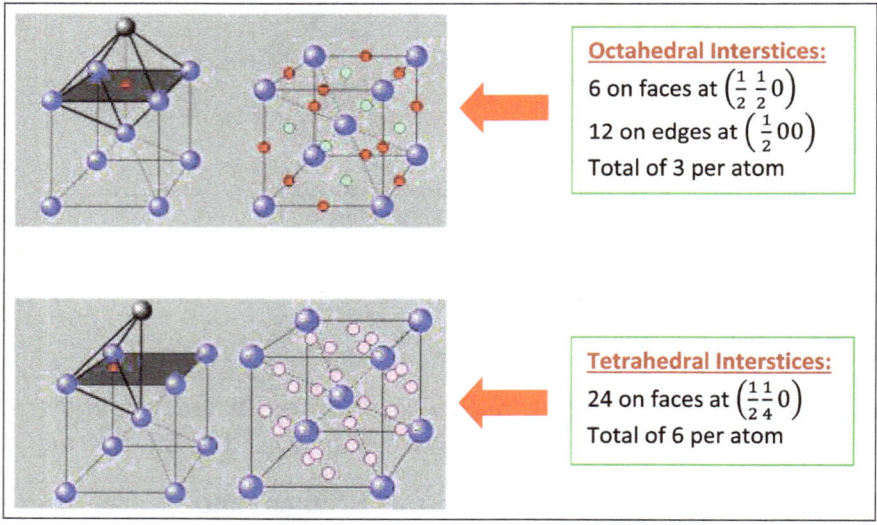

Octahedral Interstices:

6 on faces at $\left(\frac{1}{2}\frac{1}{2}0\right)$

12 on edges at $\left(\frac{1}{2}00\right)$

Total of 3 per atom

Tetrahedral Interstices:

24 on faces at $\left(\frac{1}{2}\frac{1}{4}0\right)$

Total of 6 per atom

Structure of Various Ceramics

Ceramic Glass

Ceramics with an entirely glassy structure have certain properties that are quite different from those of metals. Recall that when metal in the liquid state is cooled, a crystalline solid precipitates when the melting freezing point is reached. However, with a glassy material, as the liquid is cooled it becomes more and more viscous. There is no sharp melting or freezing point. It goes from liquid to a soft plastic solid and finally becomes hard and brittle. Because of this unique property, it can be blown into shapes, in addition to being cast, rolled, drawn and otherwise processed like a metal.

Glassy behavior is related to the atomic structure of the material. If pure silica (SiO_2) is fused together, a glass called vitreous silica is formed on cooling. The basic unit structure of this glass is the silica tetrahedron, which is composed of a single silicon atom surrounded by four equidistant oxygen atoms. The silicon atoms occupy the openings (interstitials) between the oxygen atoms and share four valence electrons with the oxygen atoms through covalent bonding. The silica atom has four valence electrons and each of the oxygen atoms has two valence electrons so the silica tetrahedron has four extra valence electrons to share with adjacent tetrahedral. The silicate structures can link together by sharing the atoms in two corners of the SiO_2 tetrahedrons, forming chain or ring structures. A network of silica tetrahedral chains form, and at high temperatures these chains easily slide past each other. As the melt cools, thermal vibrational energy decreases and the chains can-not move as easily so the structure becomes more rigid. Silica is the most important constituent of glass, but other oxides are added to change certain physical characteristics or to lower the melting point.

Ceramic Crystalline or Partially Crystalline Material

Most ceramics usually contain both metallic and nonmetallic elements with ionic or covalent bonds. Therefore, the structure the metallic atoms, the structure of the nonmetallic atoms, and the balance of charges produced by the valence electrons must be considered. As with metals, the unit cell is used in describing the atomic structure of ceramics. The cubic and the hexagonal cells are most common. Additionally, the difference in radii between the metallic and nonmetallic ions plays an important role in the arrangement of the unit cell.

In metals, the regular arrangement of atoms into densely packed planes led to the occurrence of slip under stress, which gives metal their characteristic ductility. In ceramics, brittle fracture rather than slip is common because both the arrangement of the atoms and the type of bonding is different. The fracture or cleavage planes of ceramics are the result of planes of regularly arranged atoms.

The building criteria for the crystal structure are:

- Maintain neutrality.

- Charge balance dictates chemical formula.

- Achieve closest packing.

Silicate Ceramics

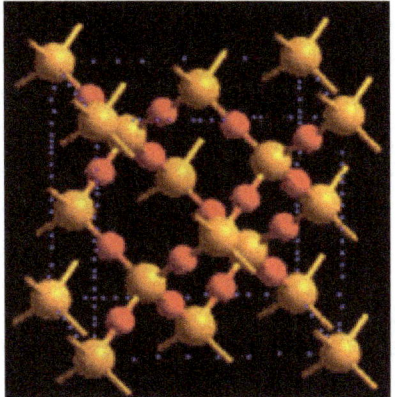

The silica structure is the basic structure for many ceramics, as well as glass. It has an internal arrangement consisting of pyramid (tetrahedral or four-sided) units. Four large oxygen (O) atoms surround each smaller silicon (Si) atom. When silica tetrahedrons share three corner atoms, they produce layered silicates (talc, kaolinite clay, mica). Clay is the basic raw material for many building products such as brick and tile. When silica tetrahedrons share four corner atoms, they produce framework silicates (quartz, tridymite). Quartz is formed when the tetrahedra in this material are arranged in a regular, orderly fashion. If silica in the molten state is cooled very slowly it crystallizes at the freezing point. But if molten silica is cooled more rapidly, the resulting solid is a disorderly arrangement which is glass.

Cement

Cement (Portland cement) is one of the main ingredients of concrete. There are a number of different grades of cement but a typical Portland cement will contain 19 to 25% SiO_2, 5 to 9% Al_2O_3, 60 to 64% CaO and 2 to 4% FeO. Cements are prepared by grinding the clays and limestone in proper proportion, firing in a kiln, and regrinding. When water is added, the minerals either decompose or combine with water, and a new phase grows throughout the mass. The reaction is solution, recrystallization, and precipitation of a silicate structure. It is usually important to control the amount of water to prevent an excess that would not be part of the structure and would weaken it. The heat of hydration (heat of reaction in the adsorption of water) in setting of the cement can be large and can cause damage in large structures.

Nitride Ceramics

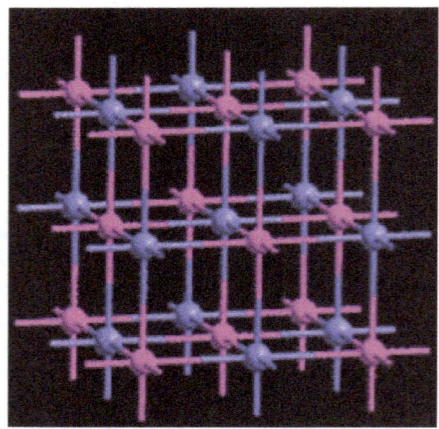

Nitrides combine the superior hardness of ceramics with high thermal and mechanical stability, making them suitable for applications as cutting tools, wear-resistant parts and structural components at high temperatures. TiN has a cubic structure which is perhaps the simplest and best known of structure types. Cations and anions both lie at the nodes of separate fcc lattices. The structure is unchanged if the Ti and N atoms (lattices) are interchanged.

Ferroelectric Ceramics

Depending on the crystal structure, in some crystal lattices, the centers of the positive and negative charges do not coincide even without the application of external electric field. In this case, it is said that there exists spontaneous polarization in the crystal. When the polarization of the dielectric can be altered by an electric field, it is called ferroelectric. A typical ceramic ferroelectric is barium titanate, $BaTiO_3$. Ferroelectric materials, especially polycrystalline ceramics, are very promising for varieties of application fields such as piezoelectric/electrostrictive transducers, and electrooptic.

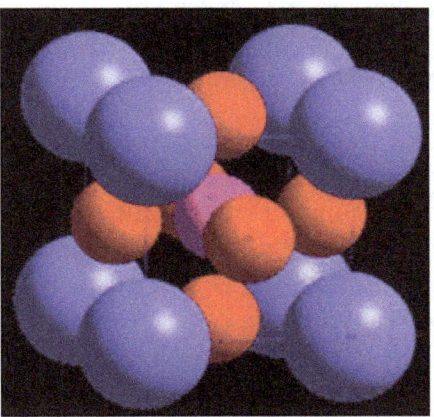

Phase Diagram

The phase diagram is important in understanding the formation and control of the microstructure of the microstructure of polyphase ceramics, just as it is with polyphase metallic materials. Also, nonequilibrium structures are even more prevalent in ceramics because the more complex crystal structures are more difficult to nucleate and to grow from the melt.

Imperfections in Ceramics

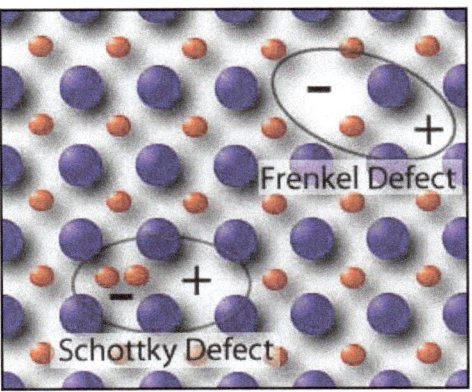

Imperfections in ceramic crystals include point defects and impurities like in metals. However, in ceramics defect formation is strongly affected by the condition of charge neutrality because the creation of areas of unbalanced charges requires an expenditure of a large amount of energy. In ionic crystals, charge neutrality often results in defects that come as pairs of ions with opposite charge or several nearby point defects in which the sum of all charges is zero. Charge neutral defects include the Frenkel and Schottky defects. A Frenkel-defect occurs when a host atom moves into a nearby interstitial position to create a vacancy-interstitial pair of cations. A Schottky-defect is a pair of nearby cation and anion vacancies. Schottky defect occurs when a host atom leaves its position and moves to the surface creating a vacancy-vacancy pair.

Sometimes, the composition may alter slightly to arrive at a more balanced atomic charge. Solids such as SiO_2, which have a well-defined chemical formula, are called stoichiometric compounds. When the composition of a solid deviates from the standard chemical formula, the resulting solid is said to be nonstoichiometric. Nonstoichiometry and the existence of point defects in a solid are often closely related. Anion vacancies are the source of the non-stoichiometry in SiO_{2-x}.

Structure of Covalent Ceramics

Most ceramic materials are neither purely covalently or ionically bonded materials. In most ionically bonded materials, there is a significant level of covalency which decreases as the difference between the electronegativities of cations and anions increases. While covalent bonding is prevalent among the group IV solids such as diamond and many other compound semiconductors, most ceramics such as NaCl, MgO, $BaTiO_3$, Fe_3O_4 etc are predominantly ionically bonded. Covalent bonding, as we saw in preceding sections, arises from the sharing of orbitals and as a result materials with this type of bonding are characterized by significant hybridization of orbitals and directionality of the bonds which play a crucial role in determining the crystal structure. In contrast, ionically bonded solids are predominantly based on the size difference between the cations and the anions and the formation of structures in them is determined by a set of rules called as Pauling's Rules.

Diamond Cubic Structure

Typical and well known purely covalent bonded materials are carbon (Diamond), Si, Ge and SiC. For example in diamond, the base lattice is FCC and is built by the C atoms with half of the tetrahedral sites filled by C atoms. Thus, the unit cell of diamond contains a total of 8 atoms. The structure is typically called as diamond cubic structure. Orbital hybridization of C atoms (sp^3) requires that the atoms are tetrahedrally coordinated and thus the structure has high degree of directionality. One unit-cell consists of two FCC motifs, one at (o o o) and another at (¼ ¼ ¼).

- What it means is that there are two FCC unit-cells of C intermingled into each other, with origin of one at (o, o, o) and another at (¼,¼,¼).

In case of other semiconducting compounds such as SiC, ZnO, ZnS, one FCC lattice can be formed by one type of atom and remaining atoms of other type occupy half of the tetrahedral sites. Figure shows the crystal structure of diamond where one can clearly observe the tetrahedrally co-ordinated C atoms.

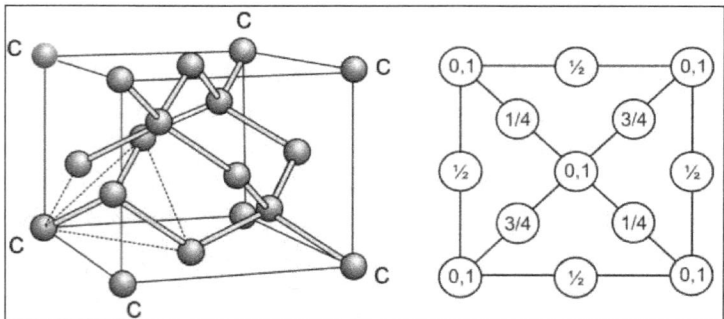

Diamond Cubic Structure (a) one unit cell showing all the positions and (b) (oo1)-plan view of the structure where positions marked show the position in zdirection while x- and positions are self-explanatory.

You can also work out the packing factor of this unit which is lower than the typical FCC unit cell. This is because the tetrahedral site size in a normal FCC unit cell is 0.225*r while in this structure, the size of the atom sitting at the site is much larger i.e. same size as the base lattice atom.

Structure of Graphite

Other forms of Carbon such as graphite and fullerene are also covalent bonded but the structures are entirely different. Graphite has a layered structure where in each layer, carbon atoms are sp^2 hybridized and they make a hexagonal pattern. However, the bonding between individual layers is van der Walls type of bonding. That is why Graphite is a soft material and is used as a lubricant.

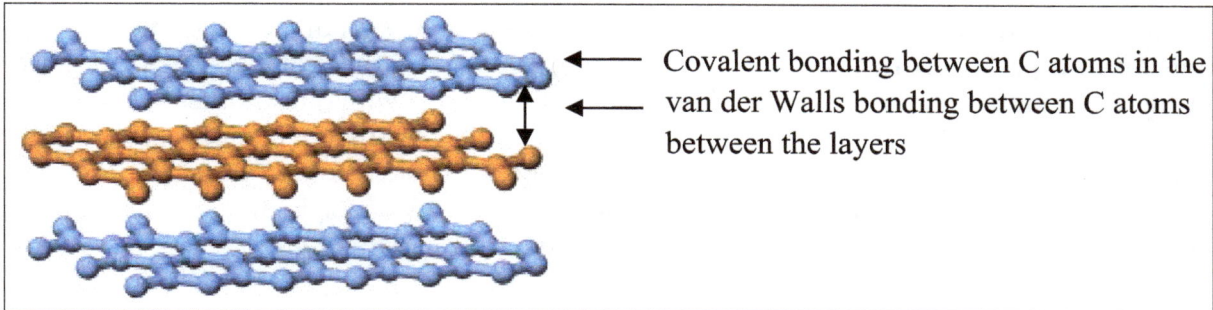

Covalent bonding between C atoms in the

van der Walls bonding between C atoms between the layers

Ionically Bonded Ceramic Structures

Most of the ceramic materials are compounds with anions and cations with different electronegativities. Hence when these ions are brought together, they form a very strong ionic bond. Typically, since anions are bigger in size than cations, anions tend to form the base lattice and cations fill in the interstices. However, it is not so simple. As there is an involvement of two different types of ions to form a crystal structure, there are certain rules or say guidelines which need to be followed to give rise to a stable crystal structure. These rules are called as Pauling's rules. Based on these rules, typically ceramic structures are based on anions forming the base lattice and cations occupying the interstices in them. Fortunately, most ceramic compounds are completely or partially ionically bonded and happen to be based on either of FCC or HCP packing of anions. As a result, we can categorize the structures of most ceramic materials into following categories:

- Compounds based on cubic closed packing (CCP or FCC) of ions.

- Compounds based on hexagonal closed packing (HCP) of ions.

- Other structures with some deviations from above two.

Pauling's Rules

Anions being the larger ions form the base lattice forming coordinated polyhedrons around cations. The co-ordination is determined by the radius ratio of cations (r_c) to anions (r_a) i.e. r_c/r_a. Also, point to note is that ionic radius of each ion is also dependent on its co-ordination.

Ligancy or Coordination number	Range of Radius Ratio (r_c/r_a)	Configuration
2	0.0-0.155	Linear
3	0.155-0.225	Triangular
4	0.225-0.424	Tetrahedral
6	0.414-0.732	Octahedral
8	0.732-1.0	Cubic
12	1.0 or above	FCC or HCP

The structure will be stable when it preserves the charge neutrality (Electrostatic valence rule). Corner linking of polyhedrons is preferred over face or edge sharing to ensure larger separation between cations. This is especially true for solids with smaller cations and cations with bigger charges e.g. Ti^{4+} and Zr^{4+}. For example, in SiO_2, due to +4 charge on Si atoms, corner linking of tetrahedrons is preferred. In a crystal containing different cations, those of high valence and small coordination number tend not to share the polyhedron elements with one another such as in materials like $BaTiO_3$. The number of essentially different kinds of constituents in a crystal tends to be small. The repeating units will tend to be identical because each atom in the structure is most stable in a specific environment. There may be two or three types of polyhedra, such as tetrahedra or octahedra, but there will not be many different types (Rule of parsimony).

Bond Strength

Bond strength is an useful parameter to determine whether the derived structure is correct or not, at least charge neutral and stoichiomteric or not. Bond strength of an ion is defined as ratio of valence of an ion to its co-ordination i.e.

$$\text{Bond strength} = \frac{\text{Valence of the ion}}{\text{Coordination number of ion}}$$

In a stoichiometric and charge neutral solid, the bond strengths of cations must be equal to those of anions. Alternatively, you can work out bond strength of one ion and from this, you can work out the valence of other ion which should match what is needed to maintain the stoichiometry and most cases, the common valence state.

Compounds based on FCC Packing of Ions

In these structures, typically anions form the FCC lattice and cations fill tetrahedral or octahedral sites in most cases, although in some cases, some other co-ordination may be preferred. Here we will discuss the following structures:

- Rocksalt structures with NaCl as parent compound.

- Fluorite and anti-fluorite structures based on CaF_2 and Na_2O.

- Zinc Blende or Sphalerite structure based on ZnS.

- Spinel structure based of formula AB_2O_4.

Rock Salt Structure

- MX Type Compounds Based on NaCl Structure.

- Anions (X) form the cation sub lattice with FCC structure.

- Cations (M) fill the octahedral sites.

- 100% occupancy of sites according to the stoichiometry since there will be one octahedral site per anion.

- Radius ratio, r_c/r_a is typically between 0.414 - 0.732 with some exceptions.

- Examples of ceramic materials with such structure as NaCl, MgO, NiO, FeO etc.

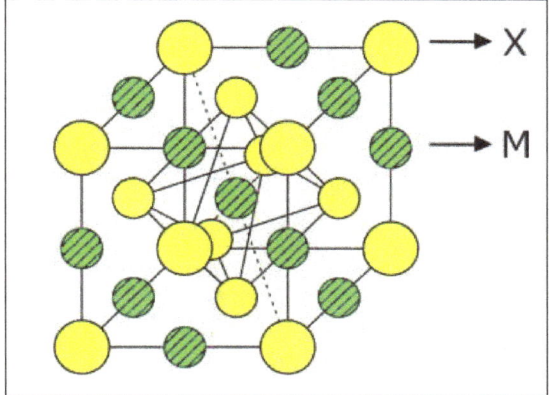

Schematic of structure of Rocksalt structured compounds.

- Lattice type: FCC and motif will be M at 0 0 0 and X at ½ 0 0.

- Four formula units per unit cell.

- Some of the radius values of cations for selected rocksalt structured compounds are given below

Compound	r_c (nm)	r_c (nm)	r_c/r_a
NaCl	0.102	0.181	0.564
MgO	0.072	0.140	0.514
SrO	0.118	0.140	0.842
NiO	0.069	0.140	0.492
FeO	0.078	0.140	0.557
MnO	0.053	0.140	0.378
PbO	0.119	0.140	0.85

- We can verify the bond-strength of ions in NaCl:

$$\text{Bond Strength of cations i.e. Na}^+ = \frac{\text{Valence of cation}}{\text{Co-ordination of cation}} = \frac{1}{6}$$

Valence of anions = Bond strength of cation × coordination of cation $= \dfrac{1}{6} \times 6 = 1.$

Which is the valence of chlorine and hence, proposed packing is appropriate.

- The octahedrons shares at the edges. If there was corner sharing of polyhedra, coordination number will be 2 for anions, which won't maintain the stoichiomentry as you can verify using bond-strength relationship.

- Schematic representation of atomic arrangement on (110) plane of rock-salt structure compounds shows the rows of empty rows of tetrahedral voids along [001]-direction.

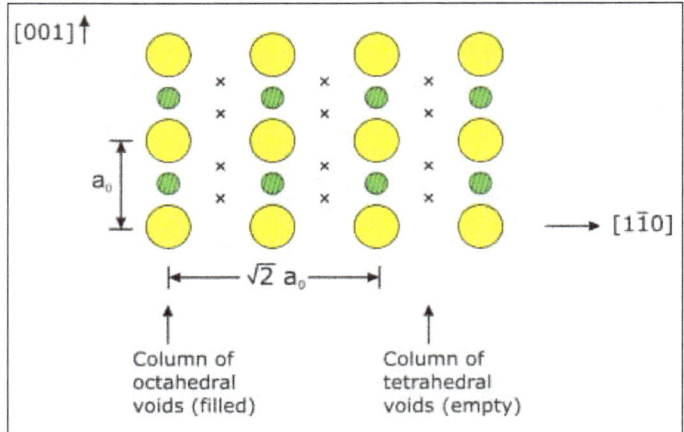

[110] plane of a Rocksalt structured compound.

You can see that if both octahedral and tetrahedral voids were filled, this would bring cations quite close to each other, leading to large electrostatic repulsion as like neighbours do not make an energetically preferred configuration. Another way of looking at this structure is to visualize hexagonal arrays of cations and anions stacked along [111]-direction repeating in an alternative fashion. Complete structure can be viewed as two FCC lattices, one of Na and another of Cl, interpenetrating into each other.

Antifluorite (A_2X) and Fluorite (AX_2) Structures

Antifluorite

- FCC packing of anions.

- All tetrahedral sites filled by cations.

- Coordination : Anions: 8, Cations: 4.

- Chemical formula: M_2X.

- Example: Li_2O, Na_2O, K_2O.

- Radius ratio (r_c/r_a): 0.225-0.414.

- Examples: $r_{Li}+$: 0.059 nm, $r_{Na}+$: 0.099 nm, $r_O{}^{2-}$: 0.14 nm.

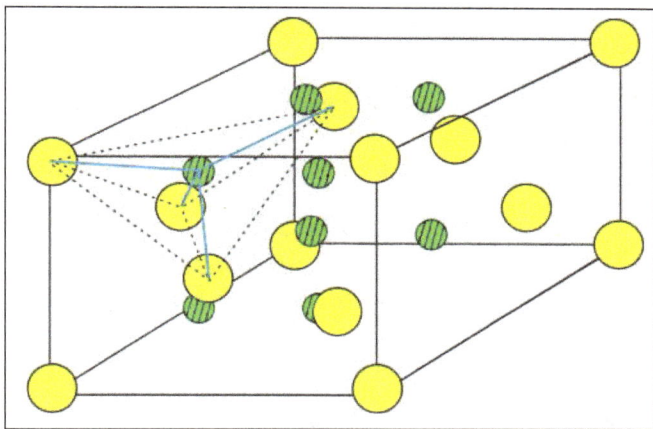

Antifluorite structure

- Lattice type: FCC.

- Motif – X: o o o, M $-\dfrac{1}{4}\dfrac{1}{4}\dfrac{1}{4}, \dfrac{3}{4}\dfrac{3}{4}\dfrac{3}{4}$.

- Four formula units per unit cell.

- In this structure in many cases, although r_c/r_a ratio predicts an octahedral co-ordination, tetrahedral coordination is preferred to fulfill the stoichiometry requirements. In turn, anions are cubic coordinated by cations (CN:8).

- The structure shows corner sharing of tetrahedra.

Fluorite Structure (CaF$_2$ Structure)

- Slightly bigger cations in comparison to other structures.

- Example: UO_2, ZrO_2, CaF_2, CeO_2.

- Typical representation of the structure appears as if cations make a FCC lattice and anions occupy the tetrahedral sites.

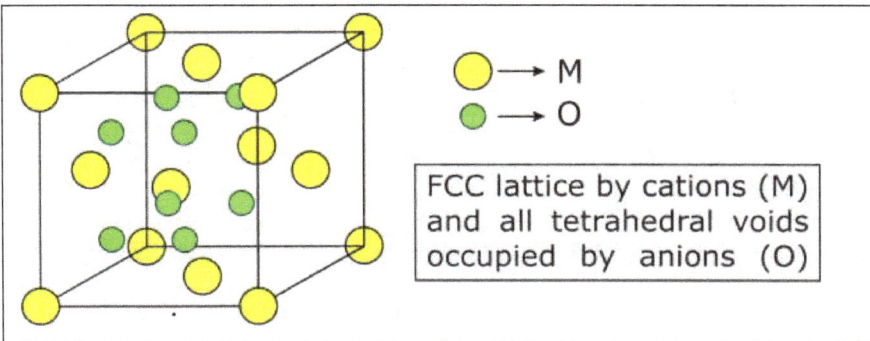

Fluorite structure

While more appropriate Fluorite structure representation is shown below where eight primitive cubic unit cells made by anions are joined together to make a big cube and cations occupy the centers of four of these small cubes in an ordered fashion.

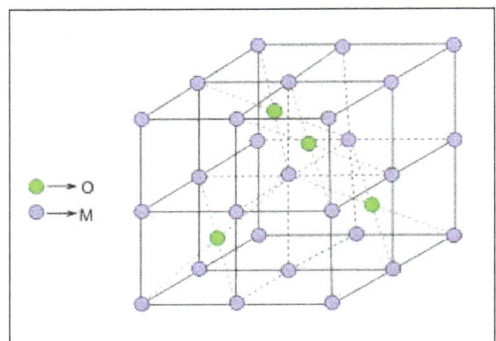

A more appropriate representation of fluorite structure.

- Co-ordination number: Cations - 8; Anions - 4.

- Lattice: FCC.

- Motif: M – 0 0 0; X – ¼ ¼ ¼; ¾ ¾ ¾ Examples of ionic radii of a few ions: U^{4+}: 0.1 nm, Zr^{4+}: 0.084 nm, Ce^{4+}: 0.097 nm, O^{2-}: 0.14 nm (observe that the cations are quite large as compared to oxygen ions) The structure as you can also see has a large void in the center of unit cell made by cations. These empty spaces make such oxides good ionic conductors which in useful in applications such as energy storage e.g. batteries.

- For having some fun with the structure, we can also draw as projection of this material on (110) plane. Here you can see the row of empty octahedral sites along [110]-direction.

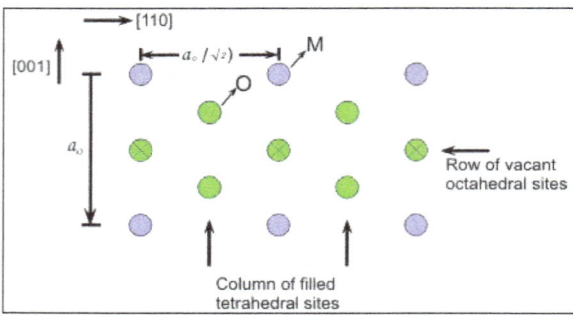

View of [110] plane of fluorite structure.

Zinc Blende (MX) Structure

- MX type compounds, also called as Sphalerite structured compounds based on mineral name of Sphalerite.

- Mostly oxides and sulphides follow this structure. Examples are ZnO, ZnS, BeO etc.

- Some covalently bonded materials and compounds have similar structure such as GaAs, SiC, BN. You can also visualize diamond also having similar structure with both anion and cation being of same type.

- Typically compounds with tetrahedral co-ordination assume this structure.

- In this structure anions form FCC lattice and cations occupy the tetrahedral interstices.

- ◦ Due to stoichiometry, half of the tetrahedral sites are filled.

- Compounds with radius ratio $\left(\dfrac{r_c}{r_a}\right)$: 0.225-0.414 follow this structure with a few exception where bonding favours a tetrahedral coordination despite unfavourable radius ratio, especially covalently bonded compounds.

  - ◦ Examples: Zn^{2+} – 0.06 nm, Be^{2+} – 0.027 nm, O^{2-} – 0.14 nm, S^{2-} – 0.184 nm.

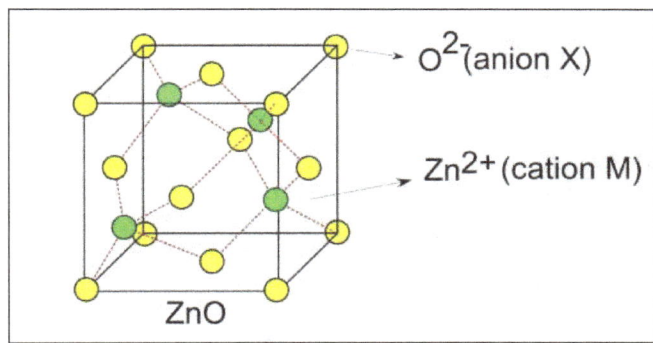

Zinc Blende or Sphalerite structure.

- Coordination numbers: M - 4, X - 4.

- Lattice type: FCC.

- Motif: M – 0 0 0; X – ¼ ¼ ¼.

- 4 formula units per unit cell.

- Tetrahedra are shared at corners.

Spinel Structure

- Formulae – $(A^{2+})(B^{3+})_2O_4$ or AB_2O_4 or $AO.B_2O_3$.

- FCC Packing of anions.

- Partial occupancy of both tetrahedral and octahedral sites i.e. 1/8th of tetrahedral and ½ of the octahedral sites are occupied.

- A spinel unit-cell is made up of eight FCC cells made by oxygen ions in the configuration $2 \times 2 \times 2$, so it is a big structure consisting of 32 oxygen atoms, 8 A atoms and 16 B atoms.

- Depending on how cations occupy different interstices, spinel structure can be Normal or Inverse.

Normal Spinel

- Chemical formula: $(A^{2+})(B^{3+})_2O_4$.

- Examples are mainly aluminates such as $MgAl_2O_4$, $FeAl_2O_4$, $CoAl_2O_4$, $NiAl_2O_4$ and few ferrites such as $ZnFe_2O_4$ and $CdFe_2O_4$.

- In this structure, all the A^{2+} ions occupy the tetrahedral sites and all the B^{3+} ions occupy the octahedral sites.

- Apply bond strength rule to verify the stoichiometry:

$$\text{Cations: } A^{2}+-\frac{2}{4}; B^{3+}-\frac{3}{6}.$$

$$\text{Oxygen valence } = \frac{2}{4} \times 1 + \frac{3}{6} \times 3 = 2$$

Inverse Spinel B(AB)O$_4$

- Chemical formula: $(A^{2+})(B^{3+}2)O4$ but can be more conveniently written as $B(AB)O_4$.

- Most ferrite follow this structure such as Fe_3O_4 (or $FeO.Fe_2O_3$), $NiFe_2O_4$, $CoFe_2O_4$ etc.

- In this structure, ½ of the B^{3+} ions occupy the tetrahedral sites and remaining ½ B^{3+} and all A^{2+} ions occupy the octahedral sites.

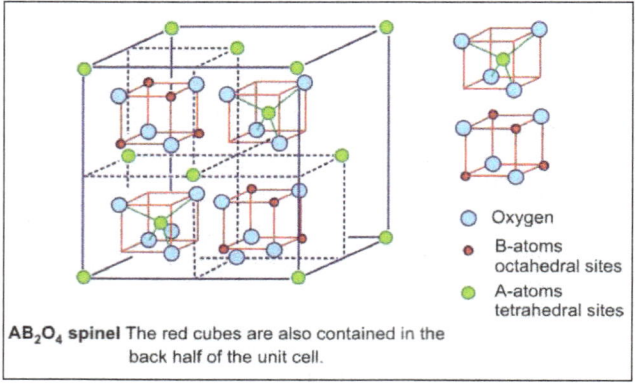

AB$_2$O$_4$ spinel The red cubes are also contained in the back half of the unit cell.

Schematic of spinel structure.

Structure of Ionic Ceramics

There are a few structures, which appear as if they are based on cubic closed packing of anions. However the actual structure is rather different and many of these structures are merely based on the cubic packing of anions. Here, we discuss the perovskite structure based on ABO$_3$ structure, CsCl structure and ReO$_3$ structure.

Perovskite (ABO$_3$) Structure

- ABO$_3$ type compounds.

- Examples are many titanates like $BaTiO_3$, $SrTiO_3$, $PbTiO_3$ etc. which happen to be technologically very useful compounds.

- In ABO$_3$ structured compounds, A ion is twelve fold coordinated by oxygen (like a dodecahedra) and B ion is octahedrally coordinated by oxygen ions.

- Oxygen atoms form a FCC-like (not FCC) cell with atoms missing from the corners which are occupied by A atoms.

- Bond strength check:

Cation Ba: $\dfrac{2}{12} = \dfrac{1}{6}$; Ti: $\dfrac{4}{6} = \dfrac{2}{3}$.

Oxygen valence $= \dfrac{1}{6} \times$ Coordination number by Ba $+ \dfrac{2}{3} \times$ coordination number by Ti $= \dfrac{1}{6} \times 4 + \dfrac{2}{3} \times 2 = 2$

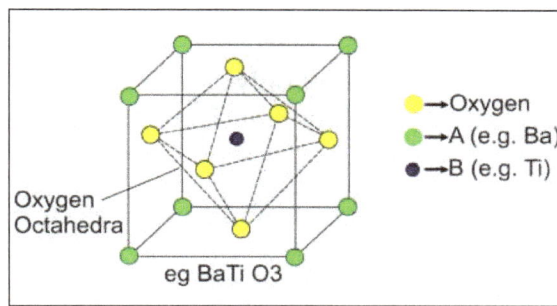

Perovskite structure.

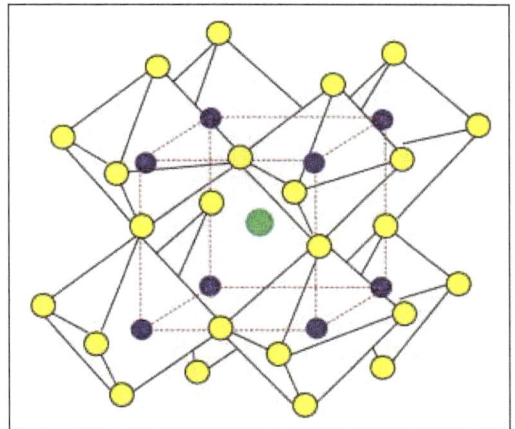

Polyhedra model of perovskite structure.

- Lattice type: Primitive Cubic.

- Motif: A ion - 0 0 0, B ion – ½ ½ ½, O ion - ½ ½ 0, 0 ½ ½, ½ 0 ½.

- One Formula unit per unit cell.

- Coordination:

  ◦ B cation is surrounded by oxygen octahedra which share corners.

  ◦ A cation is surrounded by oxygen dodecahedra which touch faces of octahedral.

- An important parameters about perovskites is the their "Tolerance Factor (t)" which is defined as,

$$t = \frac{r_A + r_O}{\sqrt{2}\left(r_B + r_O\right)}$$

- This derived from the geometry of a cube in which the atoms are of such sizes that they touch each other and hence, face diagonal of the unit cell would be $\sqrt{2}$ times the unit-cell length, as result t = 1 for a perfect cubic perovskite.

- However, due variations in the ionic radii of various ions, many perovskite show deviations from t = 1 and may not even have a cubic structure. Deviations from t = 1 signify the level of lattice distortion.

- For example $BaTiO_3$ has cubic structure only above ~120°C while it is tetragonal at room temperature and further adopts orthorhombic and rhombohedral structure if cooled below RT.

- Perovskite can also various combinations of ionic valence such as:

  ○ $A^{2+} B^{+4} O_4$ e.g. $BaTiO_3$, $PbTiO_3$, $CaTiO_3$, $SrTiO_3$ etc.

  ○ $A^{3+} B^{3+} O_4$ e.g. $LaAlO_3$, $LaGaO_3$, $BiFeO_3$.

  ○ Mixed Perovskites:

    ▪ $A^{2+}\left(B_{1/3}^{2+}B_{1/3}^{2+}\right)O_3$ e.g. $Pb\left(Mg_{1/3}Nb_{2/3}\right)O_3$.

    ▪ $A^{2+}\left(B_{1/2}^{3+}B_{1/2}^{5+}\right)O_3$ e.g. $Pb\left(Sc_{1/2}Ta_{1/2}\right)O_3$.

ReO_3 Structure

- Stoichiometry : MX3.

- Lattice type: Primitive cubic.

- Atomic Positions: M- 0 0 0; X - ½ 0 0, 0 ½ 0, 0 0 ½.

- Coordination Numbers:

  ○ M CN = 6 Octahedral coordination.

  ○ X CN = 2 Linear coordination.

- Can be visualized as perovskite ABO_3 structure with empty B-sites.

- Representative Oxides:

  ○ ReO_3, UO_3, WO_3.

- Used for gas sensing and electrochromic applications.

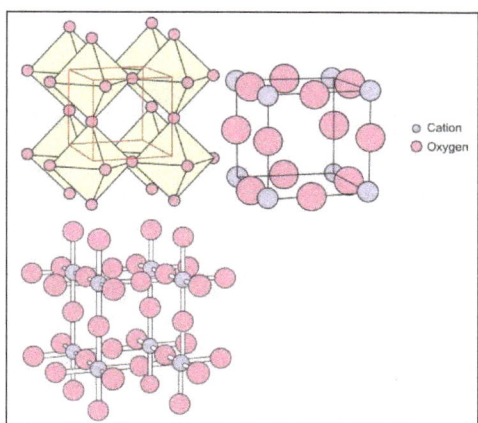

ReO$_3$ structure and polyhedra model.

CsCl Structure

- MX type compounds, parent compound being CsCl.

- Examples: Halides such as CSCl, AgI, AgBr etc.

- Radius ratio governs cubic co-ordination of both cations and anions.

- Lattice type: Primitive cubic lattice.

- Motif: Anions (X): 0 0 0, Cations (M): ½ ½ ½.

- One formula unit per unit cell.

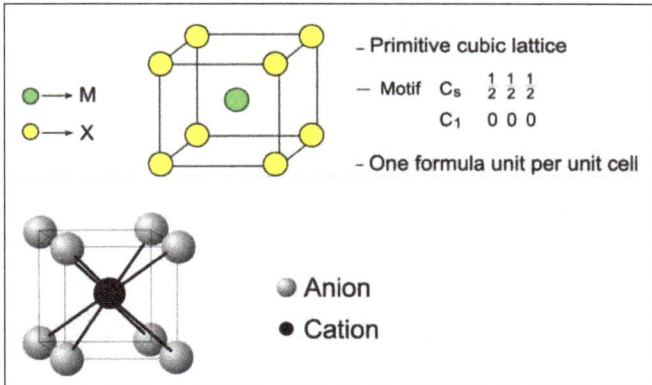

(a) CsCl structure (b) Ball-stick model.

Orthogonal Structures

Many superconductors follow the structures which contain perovskite units i.e. the structure contains the perovskite structured units stacked along c-axis or [001]-direction in most cases. The examples are superconductors such as YBa$_2$Cu$_3$O$_7$, ferroelectrics e.g. Bi$_4$Ti$_3$O$_{12}$ etc. In some other compounds such as La-Sr-Cu-O, the structure is composed of alternating perovskite and rack salt structure units. Such a way of representing these structures makes it easy to understand these complex structures. Here we will take examples of Y-Ba-Cu-O and La-Sr-Cu-O.

Yttrium Barium Copper Oxide or YBCO ($Y_3Ba_2Cu_2^{2+}Cu_3^{3+}O_7$)

Parent compound is $Y_3Cu_3^{3+}O_9$ which also contain perovskite units. Doping of Y by Ba leads to structure modification as well as reduction of Cu^{3+} too Cu^{2+} state and thus resulting in reduction in the number of required oxygen ions and hence creates oxygen vacancies in the structure. This gives a transition temperature of ~92 K below which the compound has zero electrical resistance i.e. is a superconductor.

$$Y_3Cu_3^{3+}O_9 \xrightarrow{Ba} YBa_2Cu_3^{3+}O_8 \xrightarrow{Cu^{3+}\rightarrow Cu^{2+}} YBa_2Cu_2^{2+}Cu^{3+}O_{7-x}$$

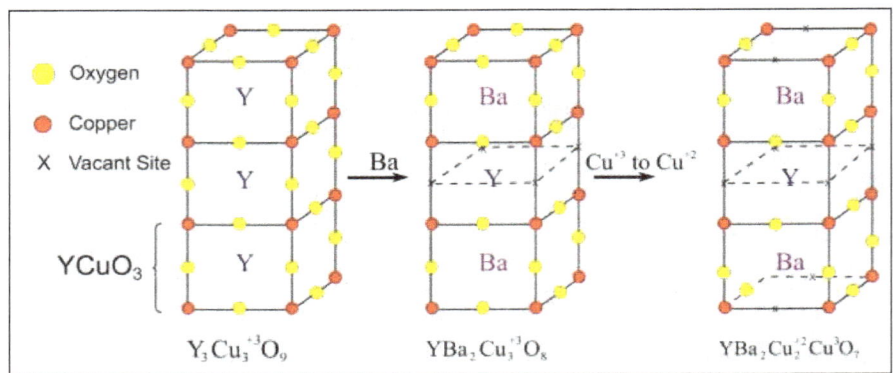

Origin of the structure of $YBa_2Cu_3O_{7-x}$ as a triple-perovskite unit.

- Here Cu coordination is of interest:
  - Cu^{2+} atoms have four-fold coordination along Cu-O chains.
  - Cu^{3+} atoms have five-fold coordination in the Cu-O planes.

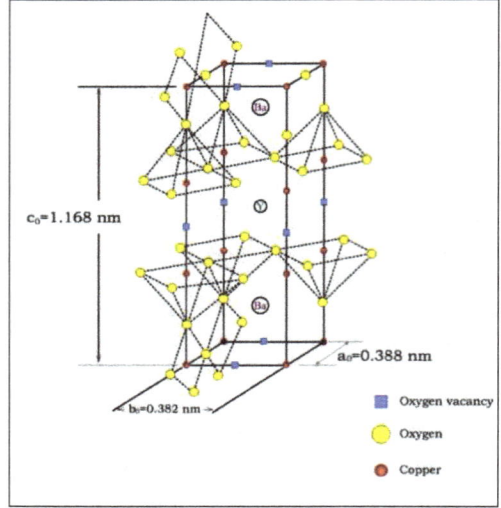

Atomic coordination in YBCO.

Lanthanum Strontium Copper Oxide ($La_{2-x}Sr_xCuO_4$)

Parent compound is La_2CuO_4 which is actually a mixture of one Rocksalt structured compound,

LaO and one perovskite structured compound, $LaCuO_3$ and can also be written as $LaO.LaCuO_3$. The structure shows a layered structure with layers stacked as $A_4O-AO_4-A_4O$ as shown below where A is La. Substitution of La by Sr results in the compound $(La2-xSrxCuO4)$ turning into a superconductor with a $T_c \sim 35K$.

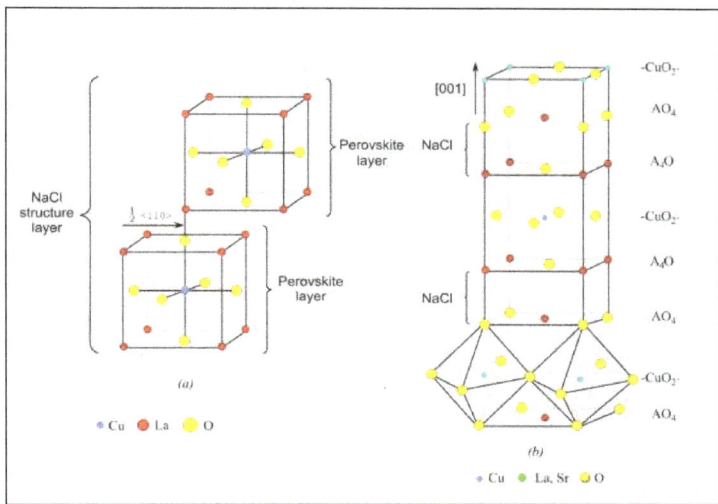

(a) Origin of $La_{2-x}Sr_xCuO_4$ structure, shown in (b), as two perovskite unit cells.

Structures based on HCP Packing of Ions

Similar to FCC packing of anions, many ceramic structures are also based on another type of closed packing of anions i.e. hexagonal closed packed (HCP). In this category we will look at the following structures:

- Wurzite structured compounds.

- Corundum structured compounds.

- Ilmenite structure compounds.

- Lithium niobate structured compounds.

- Rutile structure.

Wurtzite (MX) Structured Compounds

- Compounds with $M^{2+}X^{2-}$ stoichiometry.

- Examples are the polymorphs of Sphalerite structured compounds such as ZnS ZnO, SiC.

- Co-ordination of both anions and cations is 4, as in Sphalerite structured compounds.

- Anions form a HCP lattice with ½ of the tetrahedral sites with by cations:

  ○ The only difference to Sphalerite structure is that here anions pack in the form of AB-CABC stacking.

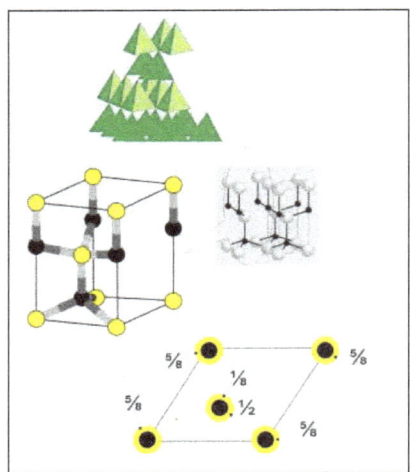

Wurtzite structure and polyhedral model.

- As you can notice, the all tetrahedrons point in one direction i.e. along the c-axis of the unit-cell and they share the corners.

- Lattice type: Primitive, HCP.

- Motif: M: o o o and $\frac{1}{2}\frac{2}{3}\frac{1}{2}$; X: o o $\frac{5}{8}$ and $\frac{1}{2}\frac{2}{3}\frac{1}{8}$.

- The filling of structure can be seen below:

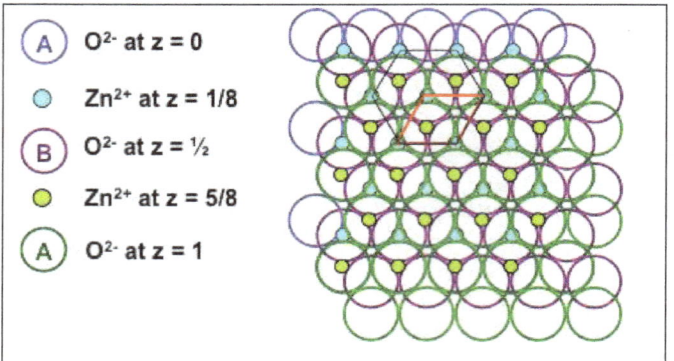

Layer by layer filling in Wurtzite.

Corundum (Al_2O_3) Structured Compounds

- M_2X_3 type of compounds.

- α-Alumina or Sapphire (Al_2O_3) is the parent compound.

  ○ other examples are compounds like Cr_2O_3, Fe_2O_3.

- Anions form a HCP lattice.

- Two-third of octahedral voids are occupied by the cations to maintain the stoichiometry.

- Co-ordination numbers: M: 6, X: 4.

- This arrangement preserves the charge neutrality as you can also verify using bond strength formula.

- This can be best viewed when we look at the basal plane of (0001)-plane of the unit-cell and start filling the interstices.

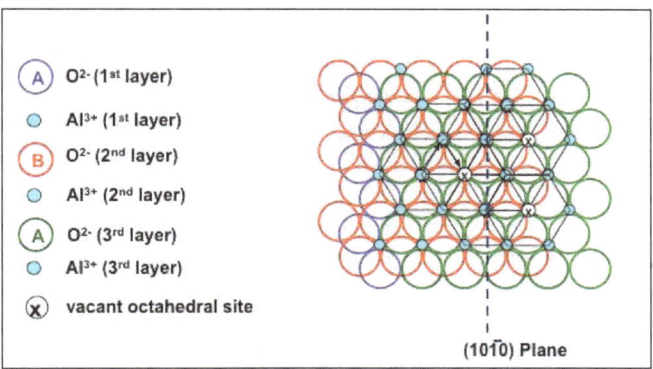

Layered filling of Corundum.

- One unit-cell consists of six layers of oxygen ions.

- A side view of the structure on [10$\bar{1}$0] plane can be seen below where you can see columns of cations along the c-axis with 2/3rd filling of octahedral sites.

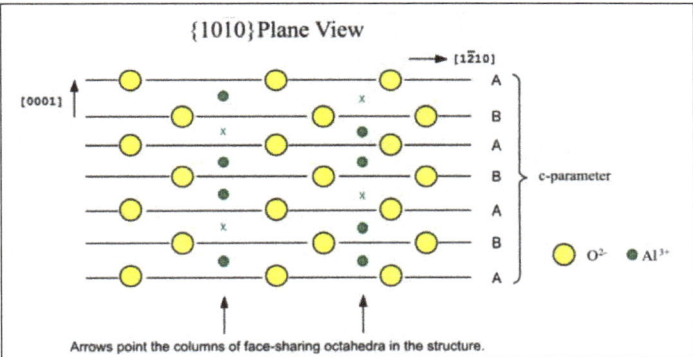

Ilmenite Structure

- The stoichiomteric formula is ABO_3 (different to perovskite ABO_3).

- The parent compound is $FeTiO_3$.

- Other compounds which follow this structure are $CdTiO_3$, $CoTiO_3$, $CrRhO_3$, $FeRhO_3$, $FeVO_3$, $LiNbO_3$, $MgGeO_3$, $MgTiO_3$.

- This structure is very similar to Corundum or α-Al_2O_3.

  - Imagine the Corundum structure and replace Al atoms in the octahedral sites in one (0001)-layer i.e. half of the total aluminum atoms by Fe and remaining half in next layer by Ti atoms in the octahedral sites and continue this order of substitution along the c-axis of the unit-cell.

- Hence, the atomic arrangement is similar to Al_2O_3 except with alternate layers of Fe and Ti in place of Al.

- Coordination numbers: both Fe and Ti remain octahedrally coordinated while O is coordinated by 4 cations i.e. 2 Fe and 2 Ti.

- Bond strength rule gives correct oxygen valence:

$$\frac{2(\text{valence of Fe})}{6(\text{CN of Fe})} \times 2(\text{CN of o by Fe}) + \frac{4(\text{valence of Ti})}{6(\text{CN of Ti})} \times (\text{CN of o by Ti}) = 2 = \text{Oxygen valence}$$

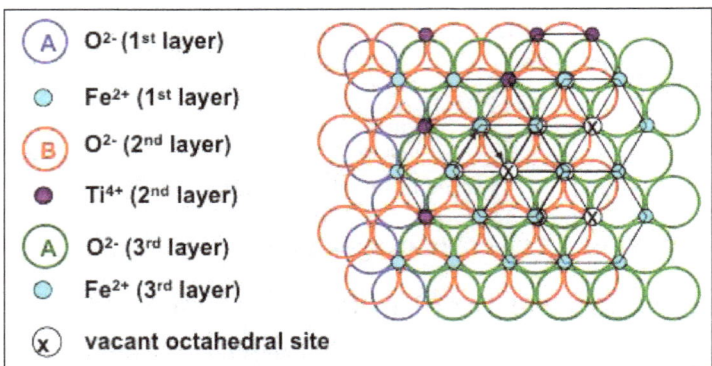

Layered filling of Ilmenite.

One unit-cell consists of six layers of oxygen ions. A side view of the structure on [$10\bar{1}0$] plane, as shown below, shows the columns of cations along the c-axis with 2/3rd filling of octahedral sites which are alternately filled by Fe and Ti ions and then followed by a vacant site.

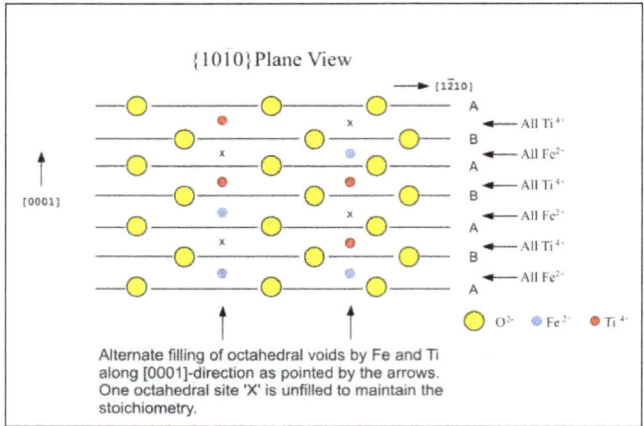

Ilmenite structure on [$10\bar{1}0$] plane.

Lithium Niobate Structure

Structure is similar to Al_2O_3 except that Al sub-lattice is substituted in an ordered manner by Li and Nb ions in the same layer unlike in alternating layer in $FeTiO_3$. The parent compound $LiNbO_3$ is ferroelectric in nature and hence, is technologically important. $LiNbO_3$ also has highly anisotropic refractive index and it shows Birefringence which is changeable by electric field. Such materials are used in electro-optic devices.

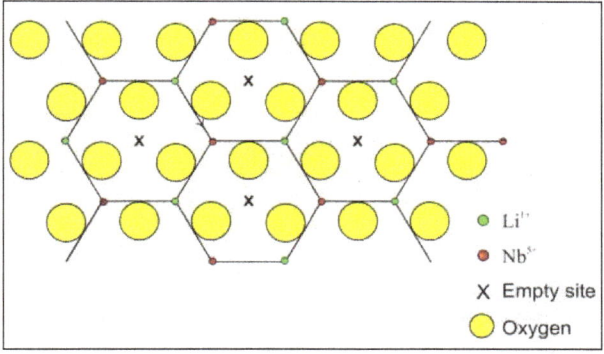

Atomic arrangement of a layer in LiNbO$_3$ structure.

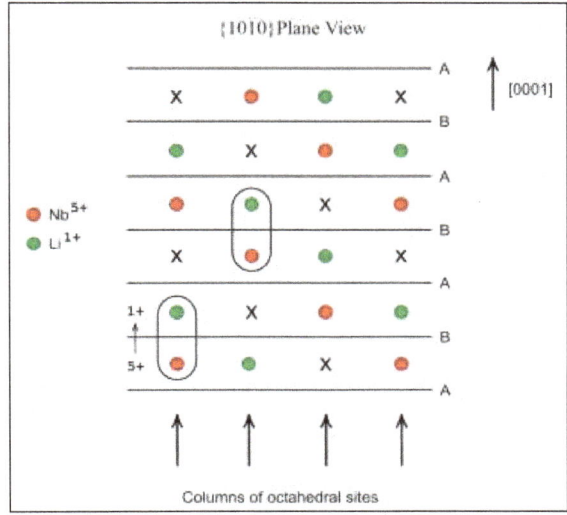

Structure on [10$\bar{1}$0] plane in LiNbO$_3$.

Rutile Structure

- Polymorph of titanium di-oxide or TiO$_2$.

  ○ Other forms are Anatase and Brookite.

- It is formed by quasi-HCP packing of anions.

- Half of the octahedral sites are filled by cations.

- Resulting structure has tetragonal crystal structure due to slight distortion in the lattice.

- Anisotropic diffusion properties of cations are found in TiO$_2$.

- Material shows large and anisotropic refractive index and high Bi-refringence.

- TiO$_2$ is often as pigments and is non-toxic.

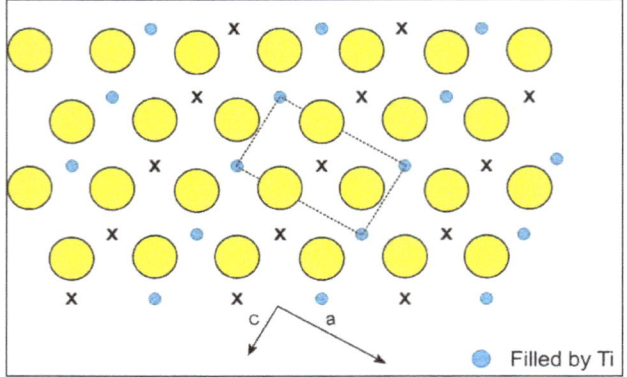

Structure of a layer of oxygen and Titanium in Rutile.

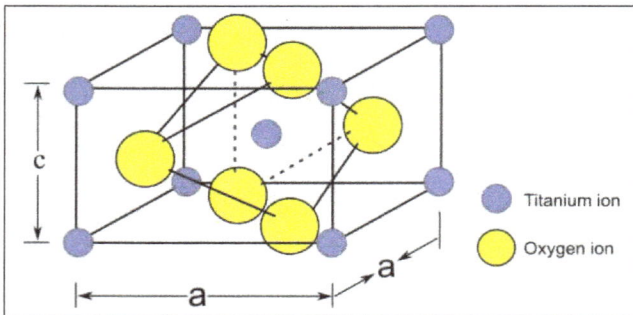

Unit-cell of Rutile.

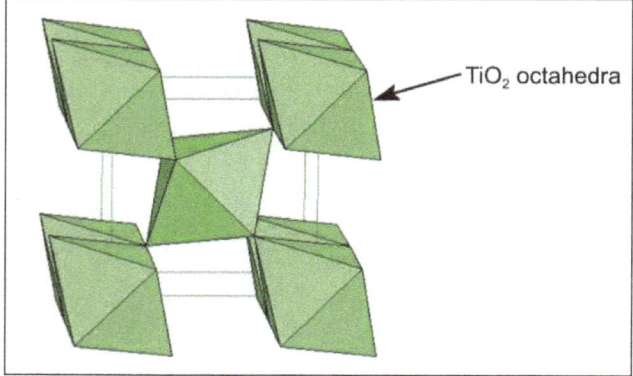

Polyhedral model of Rutile.

Particle Size Distribution

Particle size is one of the most important properties of clays and ceramic minerals. One gram of powdered material can have billions of particles with a surface area of many square meters. A knowledge of not only the average size but how the sizes are distributed can be very valuable.

The sieve analysis test is performed by slurrying and soaking a clay sample in water and then washing it through a series of successively finer and finer interlocking screens (this can be done on a plastic or lump clay material by drying and then slaking it thoroughly in water first). This test is important in monitoring consistency of clay and material powders (unlike most of the other tests, this one does require that you purchase equipment you cannot easily make).

Particle size tests are important in many industries (e.g. cement) and sieves have thus become highly standardized. Usually screens are chosen so that each has an opening with double the area of the one below it. This is referred to as a series because the width of the 'root of two' opening is the root of two (or 1.414) times that of the previous. Screens can be referenced in mesh sizes (a 100 mesh screen has about 100 wires per inch) or dimensions of the opening. They are calibrated from a standard 200 mesh having an opening of 0.0029.

Insight-live.com defines a test abbreviated 'SIEV', it clearly defines the procedure to use to perform a particle size appraisal of a material. It also provides information on the reasons why the test should be a part of your standard testing regimen.

One of the most important aspects of doing sieve analysis tests is the matter of selecting a representative specimen to test. It does little good to graph in detail an analysis if the specimen is not typical. The diagram shown here illustrates one method.

Taking a Representative Specimen

In addition to controlling material consistency and spotting impurities, sieve analysis tests are valuable in making formulation decisions. For example, the strongest and best clays are made dense by a good distribution of particle sizes at plus 325 mesh. In this way, there is a good packing in the dried matrix and therefore maximum green strength and minimum low temperature absorption. Bodies with a distribution of particle sizes tend to be comparatively stronger in the fired state and tend to have more robust working and drying properties than those made from the same materials ground to a fine mesh size. Particles of 100 mesh or finer do not generally have a noticeable texture (they feel smooth) but are easily detected by a sieve analysis.

A sieve analysis test can also alert you to trouble. It will show if coarser or agglomerated particles are present in a body presenting the danger of pinholing and premature bloating (especially in oxidation firing if the particles are "undesirables" e.g. coal, plaster, concretions). The open nature of a coarser body can also invite pinholing and pitting in less viscous glazes.

Clays with sand or grog additions in a very fine plastic particle size matrix often lack intermediate sizes and can micro-crack around each large particle. This phenomenon, although invisible to the eye, is evident not only by the dead 'thud' when a piece is tapped with a pen but by a sieve analysis of the raw material. A sieve analysis can be employed to incorporate a coarser clay or a finer grog or sand.

A Root-of-two Series of Test Sieves

The coarsest screen is at the top, the finest on the bottom. The opening for each is shown on the label. They are chosen such that each successive screen going down has an opening that is about half the area of the one above it. Using this series you can produce a practical measurement of the distribution of particle sizes in ceramic materials and bodies used in traditional ceramics (structural products industries, like brick, measure coarser particles than this, starting at perhaps 10 mesh and ending at 70). The 325 screen on the bottom is only used sometimes, it is difficult to finer-that-325 particles to pass through it because it blinds. It is not possible to shake powder through sieves that are this fine, samples must be washed through.

Particle Size Distribution and Root-of-two Stack of Sieves from 40-325 Mesh

The coarsest screen at the top has an opening of 425 microns (that means that 425 micron and finer particles will pass through it, and, conversely, plus 425 micron particles will not pass). It's opening has an area of 180,000 square microns (425 x 425). Going downward, the openings on each screen have areas half that of the one above. Thus, for the second screen down, the opening area is 90,000 microns. Standardized sieves like this are essential to the study, classification and maintenance of powdered materials.

What Labs use to Measure Particle Size?

To measure particle size in a slurry or powder you need sieves. This is the most popular type used in labs. They are made from brass by a company named Tyler. The range of screen sizes for testing particle size is very wide (obvious here: the top screen has an opening of 56 mm, the bottom one 0.1 mm - the wires are almost too small to see). The finer sieves (especially 200) are fragile and easily ripped. It is good to have a 50, 100 and 150.

Traditional Ceramics

Traditional ceramics are the ceramic materials that are derived from common, naturally occurring raw materials such as clay minerals and quartz sand. Through industrial processes that have been practiced in some form for centuries, these materials are made into such familiar products as china tableware, clay brick and tile, industrial abrasives and refractory linings, and portland cement.

Traditional ceramic objects are almost as old as the human race. Naturally occurring abrasives were undoubtedly used to sharpen primitive wood and stone tools, and fragments of useful clay vessels have been found dating from the Neolithic Period, some 10,000 years ago. Not long after the first crude clay vessels were made, people learned how to make them stronger, harder, and less permeable to fluids by burning. These advances were followed by structural clay products, including brick and tile. Clay-based bricks, strengthened and toughened with fibres such as straw, were among the earliest composite materials. Artistic uses of pottery also achieved a high degree of sophistication, especially in China, the Middle East, and the Americas.

With the advent of the Metal Age some 5,000 years ago, early smiths capitalized on the refractory nature of common quartz sand to make molds for the casting of metals—a practice still employed in modern foundries. The Greeks and Romans developed lime-mortar cements, and the Romans in particular used the material to construct remarkable civil engineering works, some of which remain standing to this day. The Industrial Revolution of the 18th and 19th centuries saw rapid improvements in the processing of ceramics, and the 20th century saw a growth in the scientific understanding of these materials. Even in the age of modern advanced ceramics, traditional ceramic products, made in large quantities by efficient, inexpensive manufacturing methods, still make up the bulk of ceramics sales worldwide. The scale of plant operations can rival those found in the metallurgical and petrochemical industries.

Raw Materials

Because of the large volumes of product involved, traditional ceramics tend to be manufactured from naturally occurring raw materials. In most cases these materials are silicates—that is, compounds based on silica (SiO_2), an oxide form of the element silicon. In fact, so common is the use of silicate minerals that traditional ceramics are often referred to as silicate ceramics and their manufacture is often called the silicate industry. Many of the silicate materials are actually unmodified or chemically modified aluminosilicates (alumina [Al_2O_3] plus silica), although silica is also used in its pure form. Altogether, the raw materials employed in traditional ceramics fall into three commonly recognized groups: clay, silica, and feldspar.

Clay

Clay minerals such as kaolinite ($Al_2[Si_2O_5][OH]_4$) are secondary geologic deposits, having been formed by the weathering of igneous rocks under the influence of water, dissolved carbon dioxide, and organic acids. The largest deposits are believed to have formed when feldspar ($KAlSi_3O_8$) was eroded from rocks such as granite and was deposited in lake beds, where it was subsequently transformed into clay.

The importance of clay minerals to traditional ceramic development and processing cannot be overemphasized. In addition to being the primary source of aluminosilicates, these minerals have layered crystal structures that result in plate-shaped particles of extremely small micrometre size. When these particles are suspended in or mixed with water, the mixture exhibits unusual rheology, or flow under pressure. This behaviour allows for such diverse processing methods as slip casting and plastic forming. Clay minerals are therefore considered to be formers, allowing the mixed ingredients to be formed into the desired shape.

Silica and Feldspar

Other constituents of traditional ceramics are silica and feldspar. Silica is a major ingredient in refractories and white wares. It is usually added as quartz sand, sandstone, or flint pebbles. The role of silica is that of a filler, used to impart "green" (that is, unfired) strength to the shaped object and to maintain that shape during firing. It also improves final properties. Feldspars are aluminosilicates that contain sodium (Na), potassium (K), or calcium (Ca). They range in composition from $NaAlSi_3O_8$ and $KAlSi_3O_8$ to $CaAl_2Si_2O_8$. Feldspars act as fluxing agents to reduce the melting temperatures of the aluminosilicate phases.

Processing

Beneficiation

Compared with other manufacturing industries, far less mineral beneficiation (*e.g.*, washing, concentrating, sizing of particulates) is employed for silicate ceramics. Clays going into common structural brick and tile are often processed directly as dug out of the ground, although there may be some blending, aging, and tempering for uniform distribution in water. Such impure clays are workable in untreated form because they already contain fillers and fluxes in association with the clay minerals. In the case of whitewares, for which the raw materials must be in a purer state, the clays are washed, and impurities are either settled out or floated off. Silicas are purified by washing and separating unwanted minerals by gravity and by magnetic and electrostatic means. Feldspars are beneficiated by flotation separation, a process in which a frothing agent is added to separate the desired material from impurities.

Blending

The calculation of amounts, weighing, and initial blending of raw materials prior to forming operations is known as batching. Batching has always constituted much of the art of the ceramic technologist. Formulas are traditionally jealously guarded secrets, involving the selection of raw materials that confer the desired working characteristics and responses to firing and that yield the sought-after character and properties. Clays must be selected on the basis of workability, fusibility, fired colour and other requirements. Silicas, likewise, must meet criteria of chemical purity and particle size distribution.

Forming

The fine, platy morphology of clay particles is used to advantage in the forming of clay-based ceramic products. Depending upon the amount of water added, clay-water bodies can be stiff or plastic. Plasticity arises by virtue of the plate-shaped clay particles slipping over one another during flow. (Nonclay ceramics can be similarly formed if plasticizers—usually polymers—are added to their mixes. In many cases organic binders are used to help hold the body together until it is fired.) With even higher water content and the addition of dispersing agents to keep the clay particles in suspension, readily flowable suspensions can be produced. These suspensions are called slips or slurries and are employed in the slip casting of clay bodies.

Plastic Forming

Plastic forming is the primary means of shaping clay-based ceramics. After the raw materials are mixed and blended into a stiff mud or plastic mix, a variety of forming techniques are employed to produce useful shapes, depending upon the ceramic involved and the type of product desired. Foremost among these techniques are pressing and extrusion.

Pressing involves the application of pressure to eliminate porosity and achieve a specific shape, depending upon the die employed. Refractory bricks, for example, are often made by die presses that are either single-action (pressing from the top only) or dual-action (simultaneously pressing from top and bottom). Structural clay products such as brick and tile can be made in the same fashion.

In pressing operations the feed material tends to have a lower water content and is referred to as a stiff mud.

The problem with die casting is that it is a piecemeal rather than a continuous process, thereby limiting throughput. Many silicate ceramics are therefore manufactured by extrusion, a process that allows a more efficient continuous production. In a commercial screw-type extruder, a screw auger continuously forces the plastic feed material through an orifice or die, resulting in simple shapes such as cylindrical rods and pipes, rectangular solid and hollow bars, and long plates. These shapes can be cut upon extrusion into shorter pieces for bricks and tiles.

Slip Casting

A different approach to the forming of clay-based ceramics is taken in slip casting of whiteware, as shown in Figure. With sufficient water content and the addition of suitable dispersing agents, clay-water mixtures can be made into suspensions called slurries or slips. These highly stable suspensions of clay particles in water arise from the careful manipulation of surface charges on the platelike clay particles. Without a dispersing agent, oppositely charged edges and surfaces of the particles would attract, leading to flocculation, a process in which groups of particles coagulate into flocs with a characteristic house-of-cards structure. Dispersing agents neutralize some of the surface charges, so that the particles can be made to repel one another and remain in suspension indefinitely. When the suspension is poured into a porous plaster mold, capillary forces suck the water into the mold from the slip and cause a steady deposition of clay particles, in dense face-to-face packing, on the inside surface of the mold. After a sufficient thickness of deposit has been obtained, the remaining slip can be poured off or drained and the mold opened to reveal a freestanding clay piece that can be dried and fired. Surprisingly complex shapes can be achieved through slip casting.

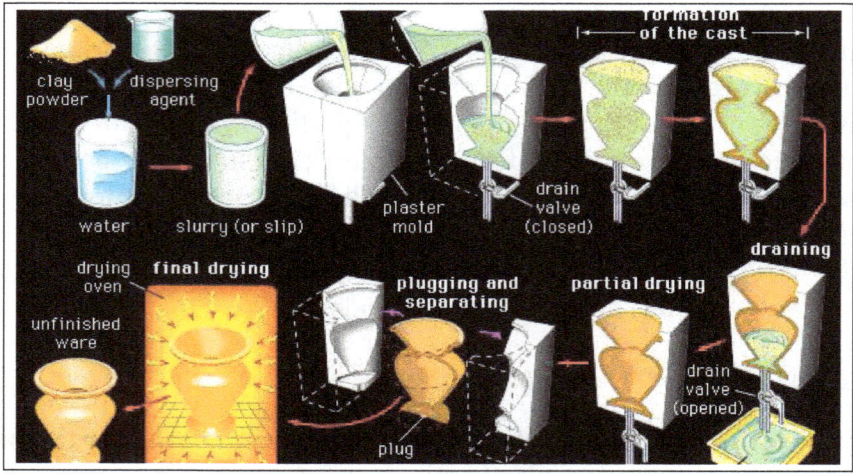

Stages in the slip casting of a thin-walled white ware container. Clay powder is mixed in water together with a dispersing agent, which keeps the clay particles suspended evenly throughout the clay-water slurry, or slip. The slip is poured into a plaster mold, where water is drawn out by capillary action and a cast is formed by the deposition of clay particles on the inner surfaces of the mold. The remaining slip is drained, and the cast is allowed to dry partially before the drain hole is plugged and the mold separated. The unfinished ware is given a final drying in an oven before it is fired into a finished product.

Firing

Kiln Operation

After careful drying to remove evaporable water, clay-based ceramics undergo gradual heating to remove structural water, to decompose and burn off any organic binders used in forming, and to achieve consolidation of the ware. Batches of specialty products, produced in smaller volumes, are cycled up and down in so-called batch furnaces. Most mass-produced traditional ceramics, on the other hand, are fired in tunnel kilns. These consist of continuous conveyor belt or railcar operations, with the ware traversing the kiln and gradually being heated from room temperature, through a hot zone, and back down to room temperature. Pyrometric cones, which deform and sag at specific temperatures, often ride with the ware to monitor the highest temperature seen in the traverse through the kiln.

Vitrification

The ultimate purpose of firing is to achieve some measure of bonding of the particles (for strength) and consolidation or reduction in porosity (e.g., for impermeability to fluids). In silicate-based ceramics, bonding and consolidation are accomplished by partial vitrification. Vitrification is the formation of glass, accomplished in this case through the melting of crystalline silicate compounds into the amorphous, noncrystalline atomic structure associated with glass. As the formed ware is heated in the kiln, the clay component turns into progressively larger amounts of glass. The partial vitrification process can be analyzed through a phase diagram such as that shown in Figure. In this diagram three crystalline phases are shown: the end members cristobalite (one crystallographic form of silica [SiO_2]) and alumina (Al_2O_3) and an intermediate compound, mullite ($3Al_2O_3 \cdot 2SiO_2$). The melting points of alumina and cristobalite, as shown on the left and right edges of the diagram, are quite high. However, intermediate compositions begin to melt at lower temperatures. As shown by the two horizontal lines on the diagram, melting begins to occur at 1,828° C (3,322° F) for high alumina compositions and as low as 1,587° C (2,889° F) for high silica compositions. (These temperatures can be lowered still further by the addition of fluxing agents, such as alkali or alkaline-earth oxide feldspars.) Between the two horizontal lines and the region of the diagram marked liquid, all compositions are only partly liquid (e.g., mullite and liquid, alumina and liquid). This partial vitrification allows for the retention of solid particles, which helps to maintain the rigidity of the ceramic piece during firing in order to minimize sagging or warpage.

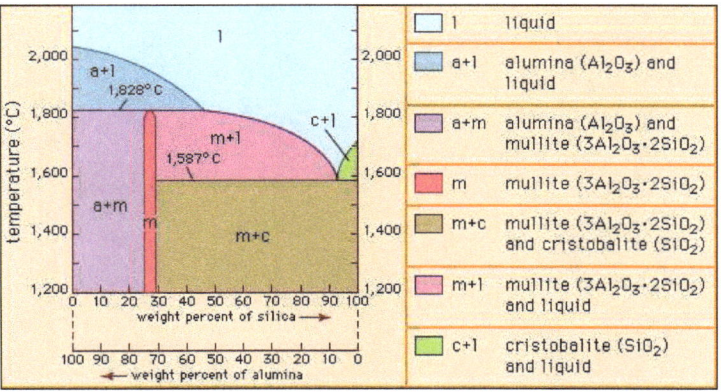

Phase diagram of the alumina-silica system. Depending on the temperature and on the content of silica and alumina, aluminosilicate clays, upon heating, form various combinations of alumina, cristobalite, mullite, and liquid. The formation of liquid phases is important in the partial vitrification of clay-based ceramics.

The role of the glassy liquid phase in the consolidation of fired clay objects is to facilitate liquid-phase or reactive-liquid sintering. In these processes the liquid first brings about a denser rearrangement of particles by viscous flow. Second, through solution-precipitation of the solid phases, small particles and surfaces of larger particles dissolve and reprecipitate at the growing "necks" that connect large particles. Rearrangement and solution-precipitation lead to bond formation and to progressive densification with reduction of porosity. A range of glass contents and residual porosities can be obtained, depending on the ingredients and the time the object is held at maximum temperature.

Finishing

If fired ceramic ware is porous and fluid impermeability is desired, or if a purely decorative finish is desired, the product can be glazed. In glazing, a glass-forming formulation is pulverized and suspended in an appropriate solvent. The fired ceramic body is dipped in or painted with the glazing slurry, and it is refired at a temperature that is lower than its initial firing temperature but high enough to vitrify the glaze formulation. Glazes can be coloured by the addition of specific transition-metal or rare-earth elements to the glaze glass or by the suspension of finely divided ceramic particles in the glaze.

Products

The raw materials and manufacturing processes outlined above produce a range of traditional ceramic products.

References

- What-are-ceramics-1769: sciencelearn.org.nz, Retrieved 08, March 2020

- Ceramic-prop, Structure, Materials: nde-ed.org, Retrieved 22, July 2020

- Particle-size-distribution-of-ceramic-powders: digitalfire.com, Retrieved 17, February 2020

- Traditional-ceramics, technology: britannica.com, Retrieved 29, May 2020

2

Ceramics: Fundamental Properties

Some of the important properties of ceramics are electrical and magnetic properties. The electrical properties that are characteristic for ceramic materials are insulating properties, electrical conductivity, dielectric strength, dielectric constant, etc. This chapter has been carefully written to provide an easy understanding of these properties of ceramics.

The properties of ceramic materials, like all materials, are dictated by the types of atoms present, the types of bonding between the atoms, and the way the atoms are packed together. This is known as the atomic scale structure. Most ceramics are made up of two or more elements. This is called a compound. For example, alumina (Al_2O_3), is a compound made up of aluminum atoms and oxygen atoms. The atoms in ceramic materials are held together by a chemical bond. The two most common chemical bonds for ceramic materials are covalent and ionic. For metals, the chemical bond is called the metallic bond. The bonding of atoms together is much stronger in covalent and ionic bonding than in metallic. That is why, generally speaking, metals are ductile and ceramics are brittle. Due to ceramic materials wide range of properties, they are used for a multitude of applications.

Electrical and Magnetic Properties of Ceramics

Although ceramics are generally considered as insulating, conducting ceramics are a very important class of materials, which are useful for various applications such as fuel cells, batteries, sensors etc. The conductivity in these materials is typically hopping type due to diffusive transport of ionic defects. Another class of electrically useful ceramics is dielectric ceramics which are insulators and are used for charge storage as memory devices or as electrical insulator only. Then, a few ceramic materials such as ferrites exhibit very good magnetic properties which make them useful for several applications such as magnetic cores, high frequency applications etc.

Diffusion and Mobility in Ceramics

Diffusion

- Diffusion causes changes in the microstructures in processes such as sintering, creep deformation, grain growth etc.

- Diffusion is also related to the transport of defects or electronic charge carriers giving rise to electrical conduction in ceramics. Ionic conductors are used in a variety of applications such as chemical and gas sensors, solid electrolytes and fuel cells. For example, an oxygen sensor made of ceramic ZrO_2 is used in automobiles to optimize the fuel/air ratio in the engines.

- Atomic diffusion rates and electrical conductivity are largely governed by defect types and their concentration where concentration is a function of temperature, partial pressure of oxygen or pO_2, and composition.

Diffusion Kinetics

Typically diffusion is explained on the basis on compositional gradients in an alloy which act as driving force for diffusion. Thermodynamically, this amounts to gradient in the chemical potential which drives the migration of species from regions of higher chemical potential to lower chemical potential so that system reaches a chemical equilibrium. The atomic flux arising as a result of driving force is expressed in terms chemical composition gradient, also called as Fick's laws.

Fick's First Law of Diffusion

It states that atomic flux, under steady-state conditions, is proportional the concentration gradient. It can be stated as:

$$\text{Atomic flux } J \propto -\left(\frac{dc}{dx}\right) \quad \text{or} \quad J = -D.\left(\frac{dc}{dx}\right)$$

where,

- J is the diffusion flux with units moles/cm²–s, and basically means amount of material passing through an unit area per unit time.

- D is the proportionality constant, called as diffusion coefficient or diffusivity in cm²/s;

- x is the position in cm.

- c is the concentration in mole/cm³.

Negative sign on the R.H.S. indicates that diffusion takes place from regions of higher concentration to the lower concentration i.e. down the concentration gradient. Diffusivity is a temperature dependent parameter and is expressed as $D = D_o \exp\left(-\frac{Q}{kT}\right)$ where Q is the activation energy, k is Boltzmann's constant and Do is the pre-exponential factor in cm²/s.

Ficks's Second Law of Diffusion

Strictly speaking, this is not a law, rather a derivation from the first law itself. It predicts how the concentration changes as a function of time under non-steady state conditions. It can be derived from Fick's first law easily.

$$\frac{dc}{dt} = -D\frac{d^2c}{dx^2}$$

where, t is time in seconds. Other terms are defined above.

Diffusivity: A Simple Model

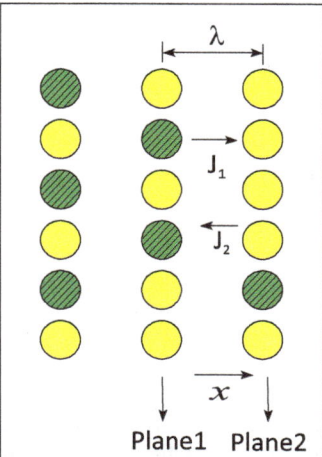

Schematic of the planes of atoms with arrows showing the cross-movement of species.

As shown in figure above a schematic diagram shows atomic planes, illustrating 1-D diffusion of species across the planes.

The flux form position $J = -D.\left(\dfrac{dc}{dx}\right)$ to $\dfrac{dc}{dt} = -D\dfrac{d^2c}{dx^2}$ is written as,

$$J_1 = \frac{1}{2}n_1\Gamma$$

Where, n_1 is the number of atoms at position (1) and Γ is the jump frequency i.e. number of atoms jumping per second (atoms/s).

Similarly,

Flux form plane $\dfrac{dc}{dt} = -D\dfrac{d^2c}{dx^2}$ to plane $J = -D.\left(\dfrac{dc}{dx}\right)$ is expressed as,

$$J_2 = \frac{1}{2}n_2\Gamma$$

Where, n_2 is the number of atoms at $\dfrac{dc}{dt} = -D\dfrac{d^2c}{dx^2}$ and 333.

In both the above expressions, factor $\dfrac{1}{2}$ is there because of equal probability of jump in $+x$ and $-x$ directions.

Now, the net flux, J, can be calculated as

$$J = J_1 - J_2 = \frac{1}{2}(n_1 - n_2)\Gamma$$

Concentration is defined as,

$$c_1 = \frac{n_1}{\lambda} \text{ and } c_2 = \frac{n_2}{\lambda}$$

If area is considered as unit area (= 1) and λ is distance between two atomic planes.

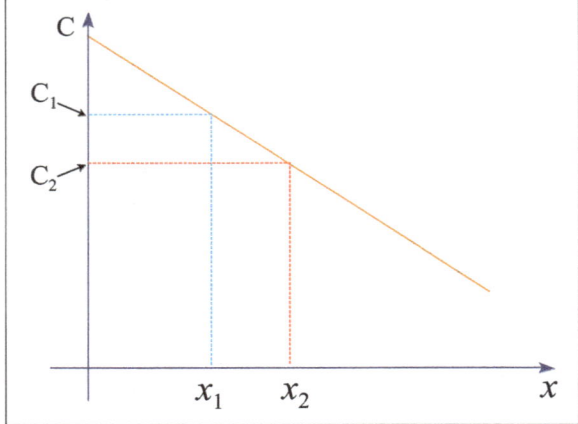

Schematic diagram showing concentration gradient between two planes of atoms Concentration gradient, as depicted in Figure, can be written as (note the minus sign).

$$-\frac{dc}{dx} = \frac{c_1}{\lambda} - \frac{c_2}{\lambda} = \frac{c_1 - c_2}{\lambda} = \frac{n_1 - n_2}{\lambda^2}$$

Hence, flux can now be expressed as,

$$J = \frac{1}{2}\left(-\lambda^2 \frac{dc}{dx}\right)\cdot\Gamma = = -\left(\frac{1}{2}\lambda^2\Gamma\right)\cdot\frac{dc}{dx}$$

$$= -D\cdot\frac{dc}{dx}$$

where, $D = \frac{1}{2}\lambda^2\Gamma$ with unit cm²/s in 1-D and can be easily shown to become $D = \frac{1}{6}\lambda^2\Gamma$ in a 3-D cubic co-ordination scenario.

In general, diffusivity can be expressed as,

$$D = \gamma\,\lambda^2\,\Gamma$$

Where, γ is governed by possible number of jumps at an instant and λ is the jump distance and is governed by the atomic configuration and crystal structure.

Temperature Dependence of Diffusivity

Now, equation $D = \gamma\,\lambda^2\,\Gamma$ can further be modified by replacing the jump frequency, Γ, which, by Boltzman statistics, is defined as,

$$\Gamma = v.\exp\left(-\frac{\Delta G^*}{kT}\right)$$

Where, v is the natural vibration frequency in s⁻¹, ΔG^* is activation energy of migration in J and k is Boltzmann Constant (J/K).

Further, ΔG^* can be written as,

$$\Delta G^* = \Delta H^* - T\Delta S^*$$

Where, ΔH^* is the enthalpy of migration and ΔS^* is the associated entropy change.

Now, substitution of equation $\Gamma = v.\exp\left(-\dfrac{\Delta G^*}{kT}\right)$ in equation $\Delta G^* = \Delta H^* - T\Delta S^*$ leads to,

$$D = \gamma.\lambda^2.v.\exp\left(\dfrac{\Delta S^*}{k}\right).\exp\left(-\dfrac{\Delta H^*}{kT}\right)$$

Or

$$D = D_o \exp\left(-\dfrac{\Delta H^*}{kT}\right)$$

Where, pre-exponential factor $D_o = \gamma\lambda^2 v^*.\exp\left(\dfrac{\Delta S^*}{kT}\right)$.

Examples of Diffusion in Ceramics

Diffusion is Lightly Doped NaCl

NaCl is often used a conducting electrolyte and it a good case for proving an example. Consider the ease of NaCl containing small amounts of $CdCl_2$ as an impurity. In such scenario, for each Cd ion, Cd occupying Na site with an extra positive charge and a sodium vacancy with one negative charge is created according to following defect reaction:

$$CdCl_2 \xrightarrow{NaCl} Ca_{Na}^{\cdot} + 2Cl_{Cl}^{x} + V_{Na}'$$

In addition, NaCl will also have certain intrinsic sodium and chlorine vacancy concentration $\left(V_{Na}' \text{ and } 2Cl_{Cl}^*\right)$ due to Schottky dissociation, depending on the temperature. In such scenario, the diffusivity of sodium ions is government by vacancy diffusion and can be worked out as,

$$D_{Na} = \left[V_{Na}'\right]\gamma\lambda^2 v \exp\left[-\dfrac{\Delta G_{Na}^*}{kT}\right]$$

Where, ΔG_{Na}^* is the migration free energy for sodium vacancies and V_{Na}' is the sodium vacancy concentration. This diffusivity of vacancy is dependent on vacancy's concentration which is related to the dopant concentration. However, the diffusivity dependence on temperature shows two regimes as shown in figure; low temperature extrinsic regime where vacancy concentration is independent of temperature and is determined by solute concentration and another high temperature intrinsic regime where vacancy concentration is dominated by intrinsic thermally degenerated vacancies.

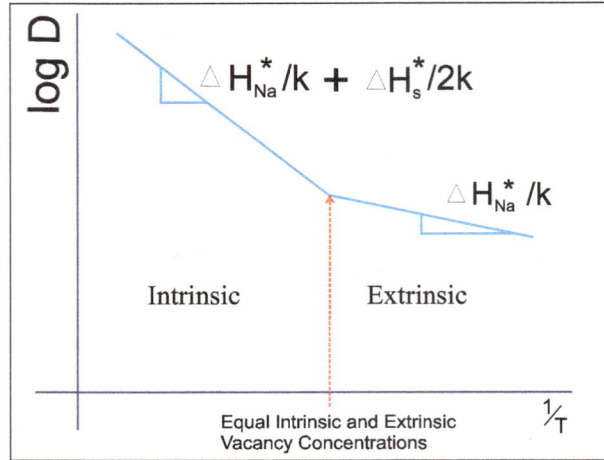

Schematic diagram showing temperature dependence of diffusivity in ceramics.

Low Temperature Regime

Extrinsic region is dominant where vacancy concentration is constant as it is determined by the solution concentration i.e. $\left[V'_{Na}\right] = \left[Cd^{\bullet}_{Na}\right]$. The diffusivity is given by,

$$D_{Na} = \left[V'_{Na}\right]\gamma\lambda^2\nu\,\exp\left[-\frac{\Delta G^*_{Na}}{kT}\right] = \left[V'_{Na}\right]\gamma\lambda^2\nu\,\exp\left[\frac{\Delta S^*_{Na}}{k}\right]\exp\left[-\frac{\Delta H^*_{Na}}{kT}\right]$$

$$= \left[V'_{Na}\right]D_o\,\exp\left[-\frac{\Delta H^*_{Na}}{kT}\right]$$

In this regime, diffusivity exhibits a temperature dependence of $-\dfrac{\Delta H^*_{Na}}{k}$.

High Temperature Regime

In this regime, the vacancy concentration is governed by intrinsic creation mechanism which Schottky defect formation and hence, diffusivity exhibits a steeper slope with higher activation energy which include not only the energy for defect migration but also for defect creation i.e.

$$\frac{\Delta H^*_{Na}}{k} + \frac{\Delta H_s}{2k}.$$

$$D_{Na} = \text{Constant} \times \exp\left(-\frac{\Delta H_s + 2\Delta H^*_{Na}}{2kT}\right)$$

At the point of crossover of two regimes, vacancy concentration due to dopants equals the thermally created vacancies.

Mobility and Diffusivity

In addition to diffusivity, one useful term which is typically used to express conduction in various ionic compounds as well as other ceramics is called as "mobility". In the context of ionic solids and ceramics, mobility is defined as velocity (v) of an entity or specie per unit driving force (F) which could be of any kind, such as chemical or electrical or mechanical. It is expressed as,

$$M = \frac{v}{F}$$

As mentioned above, the Force "F" can arise from either of chemical potential gradient or electrical potential gradient or stress gradient or interfacial energy gradient, or any other similar parameter.

Based, on these, the mobility can be defined as:

Absolute mobility,

$$B_i = \frac{v_i \left(\dfrac{cm}{s} \right)}{(1/N_A)(\partial \mu_i / \partial x)} = \frac{cm^2}{J.s}$$

where N_A is Avogadro's Number, μ_i is chemical potential in J/mol and x is the position in cm.

Chemical mobility,

$$B_i' = \frac{V_i}{(\partial \mu_i / \partial x)} = \left(\frac{mole.cm^2}{J.s} \right)$$

Electrical mobility (in the context of ceramics),

$$\mu_i = \frac{B_i}{\left(\dfrac{\partial \phi}{\partial x} \right)} = \left(\frac{cm/s}{V/s} \right)$$

For atomic transport, it was Einstein who first pointed out that the most general driving force is the virtual force that acts on a diffusing atom or species and is due to negative gradient of the chemical potential or partial molar free energy. It is expressed as,

$$F_i (J/cm) = -\frac{1}{N_a} \left(\frac{d\mu_i}{dx} \right)$$

where μ_i is the chemical potential of i and N_a is the Avogadro 's number.

Absolute mobility, B_i, is given by

$$B_i = \frac{v(cm/s)}{F(J/cm)} = -\frac{v_i}{\left[\dfrac{1}{N_A} \cdot \left(\dfrac{d\mu_i}{dx} \right) \right]} cm^2\Big/V.s$$

To obtain the relation between mobility and diffusivity of species, i, we need to write the flux in a general form as a product of concentration, c_i, and velocity, v_i, i.e.

$$J_i = c_i . v_i = c_i \, B_i \, F_i$$

Now, substituting for F_i from equation $F_i \left(J/cm \right) = -\dfrac{1}{N_a} \left(\dfrac{d\mu_i}{dx} \right)$, we have,

$$J_i = -\frac{1}{N_A} . \left(\frac{d\mu_i}{dx} \right) B_i c_i$$

Now, for an ideal solution with unit activity of species i, chemical potential can be expressed as,

$$\mu_i = \mu_i^\circ + RT \, In \, C_i$$

where, R is the gas constant. So, the change in the chemical potential can be written as,

$$d\mu_i = RT.d \, In \, c_i = \frac{RT}{c_i} In \, c_i$$

i.e.

$$\frac{d\mu_i}{dx} = \frac{RT}{c_i} \left(\frac{dc_i}{dx} \right)$$

Substituting $\dfrac{d\mu_i}{dx} = \dfrac{RT}{c_i} \left(\dfrac{dc_i}{dx} \right)$ into equation $J_i = -\dfrac{1}{N_A} . \left(\dfrac{d\mu_i}{dx} \right) B_i c_i$ leads to,

$$J_i = -\frac{RT}{N_A} . B_i . \frac{dc_i}{dx}$$

If we compare above equation with Fick's first law i.e. equation $J = -D. \left(\dfrac{dc}{dx} \right)$, diffusivity of i, D_i can be written as,

$$D_i = \frac{RT}{N_A} . B_i = kTB_i$$

Above equation is called as Nernst Einstein Equation.

Creep in Ceramics

A good understanding of the creep or time dependent plastic deformation of materials is an important requirement for advanced structural applications at elevated temperatures. Structural ceramics have considerable potential for use at high temperatures because of their high melting points and oxidation resistance. In addition, the inherent brittleness of ceramics may be partially overcome by enhancing the toughness through techniques such as transformation toughening, as in zirconia ceramics, or by incorporating whisker reinforcements, as in SiC whisker reinforced alumina.

Theoretical Models for Creep

The high temperature creep of crystalline materials may be represented by a relationship of the following form,

$$\dot{\varepsilon} = A \frac{DGb}{kT} \left(\frac{b}{d} \right)^{p} \left(\frac{\sigma}{G} \right)^{n}$$

where, $\dot{\varepsilon}$ is the steady state creep rate, A is a dimensionless constant, D is the appropriate diffusion coefficient, G is the shear modulus, b is the magnitude of the Burgers vector, k is Boltzmann's constant, T is the absolute temperature, d is the grain size, σ is the imposed stress, and p and n are constants termed the inverse grain size exponent and the stress exponent, respectively. The diffusion coefficient D may be expressed as $D_{0} \exp (-Q/RT)$, where D_{0} is a frequency factor, Q is the appropriate activation energy, and R is the gas constant (8·31 J mo1^{-1} K^{-1}). In general, several mechanisms often contribute to the creep deformation at elevated temperatures, but creep is usually controlled by only one of these mechanisms. Experimental studies of creep typically lead to a determination of the values for n, p, and Q, and an attempt to identify the possible rate controlling mechanism by comparing these values with theoretical predictions.

The theoretical models developed for creep deformation may be divided into two broad categories: boundary mechanisms, which rely on the presence of grain boundaries and occur only in polycrystalline materials, and lattice mechanisms, which are independent of the presence of grain boundaries and occur both in single crystal and polycrystalline materials. Clearly, boundary mechanisms are associated with some dependence on grain size so that $p \geq 1$ whereas lattice mechanisms occur within the grain interiors and $p = 0$. It is important to note that most of the theoretical models were developed originally for metallic materials but they apply also to ceramics.

Boundary Mechanisms

The relevant boundary mechanisms in terms of the predicted values for n, p, and Q are summarised in Table: the terms Q_i, Q_{gb}, Q_{ph} and Q_s, refer to the activation energies for lattice diffusion, grain boundary diffusion, diffusion through the viscous grain boundary phase, and a solute drag process, respectively.

The boundary mechanisms may be subdivided into three distinct categories: Lifshitz sliding, Rachinger sliding, and. interface reaction control. Grain boundary sliding is an important consequence of all boundary mechanisms, but Lifshitz sliding requires that the grains elongate along the tensile axis and maintain their adjacent neighbours whereas Rachinger sliding requires that the grains maintain essentially their original shape but exchange their neighbours. Interface reaction control becomes important when the boundaries are not perfect sinks and sources for vacancies.

Table: Creep parameters for boundary mechanisms.

Mechanism	n	p	Q
Lifshitz sliding			
Sliding accommodated by diffusion			
Nabarro- Herring creep	1	2	Q_1

Coble creep	1	3	Q_{gb}
Sliding accommodated by	1	1	Q_{gb}
Intragranular flow across grains			
Rachinger sliding			
Without glassy phase			
sliding accommodated by formation of grain boundary cavities	2	1	Q_1
sliding accommodated by triple point fold formation	3·5	2	Q_1
With glassy phase	1	1	Q_{ph}
	1	2	Q_{ph}
	1	3	Q_{ph}
Interface reaction control			
Without glassy phase	> 1	1	...
	> 1	1	...
	2	1	Q_1
	2	2	Q_{gb}
By solute drag	2	1	Q_s
With glassy phase	1	1	...

As noted in Table, Lifshitz sliding may be accommodated either by diffusional vacancy flow or by the intragranular flow of dislocations across the grains. Lifshitz sliding associated with vacancy flow is referred to as diffusion creep and the driving force for vacancy diffusion is the difference in the vacancy concentrations between the different grain boundaries. When vacancy flow occurs through the lattice the process is termed Nabarro- Herring creep and when vacancy flow occurs along the grain boundaries the process is termed Coble creep. It is important to note that these processes may be considered either as diffusion creep (grain elongation) accommodated by grain boundary sliding or as grain boundary sliding accompanied by diffusion creep. The processes of diffusion creep are understood reasonably well in metallic materials and these concepts may be extended to ceramics by considering the effect of ambipolar diffusion and the transport of cations and anions along parallel but different diffusion paths.

Rachinger sliding, in which the grains maintain their original shapes, has been examined for boundaries both without and with a glassy phase. For materials without a glassy phase, grain boundary sliding may be accommodated either by the formation of cavities or by the formation of triple point folds and both of these mechanisms give n > I. For materials where the boundaries contain a glassy phase, all the models lead to a stress exponent of n = I.

The mechanisms of Lifschitz and Rachinger sliding assume that the grain boundaries act as perfect sources and sinks for vacancies. When the grain boundaries are not perfect sources and sinks, the process of creating or annihilating vacancies may control the creep deformation. This is termed interface reaction controlled creep and, depending on the details of the theories, the mechanisms predict various values for n and p.

Lattice Mechanisms

The major lattice mechanisms of relevance to the creep of ceramics are summarised in Table: Q_{ci}

and Q_p are the activation energies for chemical inter diffusion and pipe diffusion, respectively, and each mechanism occurs intragranularly so that p = 0. Mechanisms having n > 1 are generally referred to as power law creep.

An extensive analys of creep data for ceramics has shown that, in the regime of power law creep, the results tend to cluster around stress exponents of either 5 or 3. A similar tendency has been noted also in metals, where the exponents are usually interpreted in terms of control by climb or gl ide with exponents of~ 5 and ~ 3, respectively. 35 ft is important to note that, with the exception of Harper Dorn creep, which has been suggested as a deformation mechanism in some ceramics, the mechanisms involving the intragranular motion of dislocations lead to stress exponents of ≥ 3.

Comparison of Experimental Data with Theoretical Models

Here, specific examples are examined for creep behaviour with stress exponents of 5, 3, and I, respectively.

Ceramics Exhibiting n ≈ 5

A stress exponent close to 5 in metals is generally attributed to intragranular dislocation creep controlled by the climb of dislocations at dislocation pile-ups. The mechanism proposed for this process by Weertman leads to the following steady state strain rate assuming that the piled-up arrays of dislocations decompose into groups of dislocation dipoles.

$$\dot{\varepsilon} = \frac{0.14}{b^{1.5} M^{0.5}} \frac{D_1 Gb}{kT} \left(\frac{\sigma}{G} \right)^{4.5}$$

where, M is the concentration of active dislocation sources.

In ceramics, it is necessary to consider the diffusion of the two ionic species, the cations and anions, and equation $\dot{\varepsilon} = \frac{0.14}{b^{1.5} M^{0.5}} \frac{D_1 Gb}{kT} \left(\frac{\sigma}{G} \right)^{4.5}$ can be used to describe power law creep in ceramics with D_1 replaced by the lattice diffusion coefficient for the slower moving ion.

Power law creep with n ≈ 5 has been reported at high stress levels in several polycrystalline ceramics such as KBr, KCl, LiF, NaCl, Nio, reaction sintered (RS) SiC, ThO_2, UC, and UO_2. Some of these experimental results are shown in Figure as a plot of normalised strain rate ekT/DGb versus normalised stress σ/G using the diffusion coefficient for the slower moving species through the lattice: Inspection of Figure reveals that most of the experimental data fall on lines having slope n ≈ 5. In addition, the data points for NaCl, LiF, and KCl demonstrate that materials with significantly different grain sizes give essentially identical strain rates, thereby confirming that p = 0. It is not possible to make a direct comparison between the experimental data and the creep rates predicted by equation $\dot{\varepsilon} = \frac{0.14}{b^{1.5} M^{0.5}} \frac{D_1 Gb}{kT} \left(\frac{\sigma}{G} \right)^{4.5}$ because of uncertainties in the precise value of M.

Ceramics Exhibiting n ≈ 3

There are two distinct possibilities for the rate controlling mechanism when n ≈ 3: control by dislocation glide and solute drag in a process involving glide and climb and control by climb of dislocations from Bardeen- Herring sources.

For metallic solid solution alloys, the occurrence of intragranular dislocation creep is often attributed to the sequential operation of glide and climb processes. When solute drag reduces the rate of glide, dislocation glide becomes rate controlling and n ≈ 3. For the viscous dragging of solute atom atmospheres around dislocations, the steady state creep rate is given by,

$$\dot{\varepsilon} = \frac{B}{e^2 c} \left(\frac{kT}{Gb^3} \right)^2 \left(\frac{\widetilde{D}Gb}{kT} \right) \left(\frac{\sigma}{G} \right)^3$$

where, e is the solute-solvent size difference c is the concentration of the solute, $\widetilde{D}$ is the solute inter diffusion coefficient, and 8 is a constant varying from~ 0·1 25 to 0·35 . This mechanism has been applied successfully to the creep deformation of several metallic solid solutions but it is probably not important in ceramics because most of the materials exhibiting n ≈ 3 contain few or no solutes which are likely to interfere with dislocation motion.

Table: Creep parameters for relevant lattice mechanisms (with p = o).

Mechanism	n	Q
Dislocation glide and climb controlled by climb	4·5	Q_1
Dislocation glide and climb controlled by glide	3	Q_{ci}
Dislocation climb from Bardeen- Herring sources controlled by lattice diffusion	3	Q_1
Controlled by pipe diffusion	5	Q_p
Harper- Dorn creep	1	Q_1

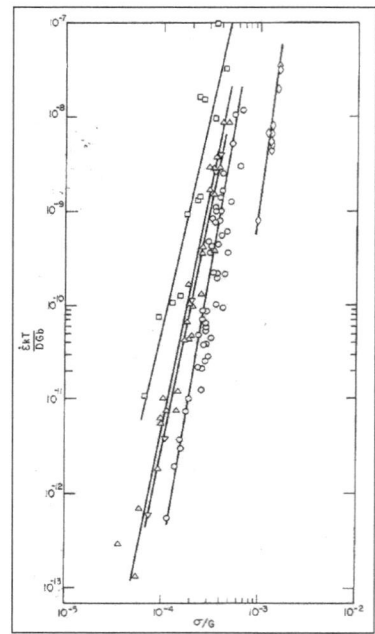

Normalised strain rate versus normalised stress for various polycrystalline ceramics showing n ≈ 5.

Ceramic		T, K	d, μm
KCl	▽	873	100-300
LiF	○	673-823	160-3000
NaCl	△	638-1015	200-3000
RS-SiC	◇	1848-1923	10; 100
UC	□	1673-1973	200

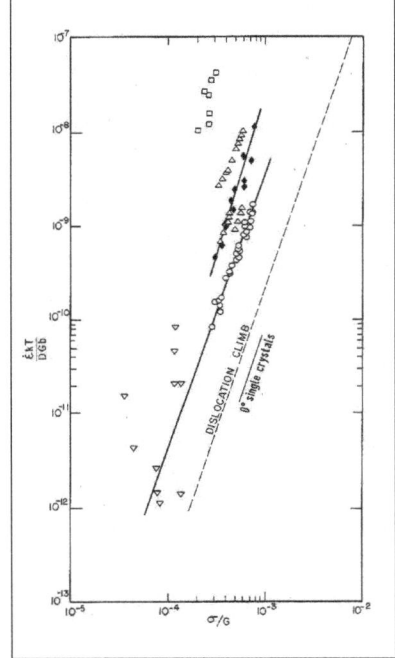

Normalised strain rate versus normalised stress for Al_2O_3 and Fe_2O_3 showing n ≈ 3.

Ceramic		T,K	d, μm
Al_2O_3	▽	1883-2120	50-100
	△	1803	13
	□	1873	9
	○	1873-1973	65
Fe_2O_3	◆	1173-1343	20

Nabarro developed a theory for creep deformation where all the creep strain is due to the climb of edge dislocations. By assuming that the climb of dislocations from Bardeen-Herring sources controls the deformation, the steady state creep rate is given by,

$$\dot{\varepsilon} = 0.22 \frac{D_1 Gb}{kT} \left(\frac{\sigma}{G} \right)^3$$

where, D_1 represents the diffusion coefficient for the slower moving species in the crystal lattice.

Creep behaviour with a stress exponent of n $\approx$ 3 has been observed in several polycrystalline ceramics such as Al_2O_3, BeO, Fe_2O_3, MgO, chemically vapour deposited (CVD) SiC, and ZrO_2 containing 10% Y_2O_3. In Figure, the experimental data for Al_2O_3 and Fe_2O_3 are shown plotted in the form of normalised strain rate versus normalised stress: again, the values of the diffusion coefficients for the slower moving species and the shear moduli. In contrast to equation $\dot\varepsilon = \dfrac{0.14}{b^{1.5}\, M^{0.5}} \dfrac{D_1 Gb}{kT}\left(\dfrac{\sigma}{G}\right)^{4.5}$, all the terms in equation $\dot\varepsilon = 0.22 \dfrac{D_1 Gb}{kT}\left(\dfrac{\sigma}{G}\right)^3$ are known and this permits direct comparison of the theoretical predictions with the experimental results. The prediction for dislocation climb through equation $\dot\varepsilon = 0.22 \dfrac{D_1 Gb}{kT}\left(\dfrac{\sigma}{G}\right)^3$ is shown as a broken line in figure. This line is in reasonable agreement with the experimental data for polycrystals but, more important, it is in excellent agreement with the experimental data for the creep of Al_2O_3 single crystals with a 0° orientation such that the stress axis is perpendicular to the basal (0001) planes to suppress basal and prismatic slip. The experimental scatter in the polycrystalline data shown in Figure is probably a result of differences in the impurity levels in the various investigations.

Ceramics Exhibiting n $\approx$ 1

In ceramics without a glassy intergranular phase, a stress exponent of n $\approx$ 1 is usually attributed to diffusion creep. When the diffusional mass transport occurs through the matrix, as in Nabarro-Herring creep, the steady state creep rate is given by,

$$\dot\varepsilon = 9.3 \frac{D_1 Gb}{kT}\left(\frac{b}{d}\right)^2 \left(\frac{\sigma}{G}\right)$$

Alternatively, when diffusional mass transport occurs along the grain boundaries, as in Coble creep, the steady state creep rate is given by,

$$\dot\varepsilon = 33.4 \frac{D_{gb} Gb}{kT}\left(\frac{\delta}{d}\right)\left(\frac{b}{d}\right)^3 \left(\frac{\sigma}{G}\right)$$

where, δ is the effective grain boundary width for diffusion.

The Nabarro-Herring and Coble creep mechanisms operate independently so that the total diffusional creep rate is given by the sum of the individual creep rates,

$$\dot\varepsilon = 9.3 \frac{D_{eff} Gb}{kT}\left(\frac{b}{d}\right)^3 \left(\frac{\sigma}{G}\right)$$

where, the effective diffusion coefficient D_{eff} is expressed as,

$$D_{eff} = D_1 + 3.6\left(\frac{\delta}{d}\right)D_{gb}$$

In ceramics, diffusion creep involves the transport of both the anions and the cations in the appropriate stoichiometric ratio. In this situation, there are two ionic species and two possible diffusion paths (lattice and grain boundary). Equation $\dot{\varepsilon} = 9.3 \dfrac{D_{eff} Gb}{kT} \left(\dfrac{b}{d}\right)^3 \left(\dfrac{\sigma}{G}\right)$ may be modified to account for the four possible rate controlling mechanisms by expressing the creep rate as,

$$\dot{\varepsilon} = 9.3 \frac{D_{com} Gb}{kT} \left(\frac{b}{d}\right)^2 \left(\frac{\sigma}{G}\right)$$

where, D_{com} is a complex diffusion coefficient defined as,

$$D_{com} = \frac{(1/\propto) D_{c(1)} + 3.6 D_{c(gb)} (\delta_e / d)}{1 + \left(\dfrac{\beta}{\propto}\right)\left(\dfrac{D_{c(1)} + 3.6 D_{c(gb)} (\delta_c / d)}{D_{a(1)} + 3.6 D_{a(gb)} (\delta_a / d)}\right)}$$

where, the subscripts c, a, I, and gb denote cation, anion, lattice, and grain boundary, respectively, and a and α a β respectively the valences of the anions and cations in an $A_\alpha B_\beta$ ceramic.

An important consequence of equation above is that creep is controlled by the faster diffusing species moving along the slower diffusion path. Owing to the large differences in the respective diffusion coefficients, equations $\dot{\varepsilon} = 9.3 \dfrac{D_{com} Gb}{kT} \left(\dfrac{b}{d}\right)^2 \left(\dfrac{\sigma}{G}\right)$ and $D_{com} = \dfrac{(1/\propto) D_{c(1)} + 3.6 D_{c(gb)} (\delta_e / d)}{1 + \left(\dfrac{\beta}{\propto}\right)\left(\dfrac{D_{c(1)} + 3.6 D_{c(gb)} (\delta_c / d)}{D_{a(1)} + 3.6 D_{a(gb)} (\delta_a / d)}\right)}$

become simplified in practice so that transitions in the mechanisms controlling creep deformation can be predicted as a simple function of grain size and temperature.4 Deformation mechanism maps have been developed to depict these transitions and recently they were extended to a consideration of diffusion creep in multicomponent ceramics of the type $A_\alpha B_\beta C_\gamma$.

In contrast to metals, because the grain sizes in ceramics are small diffusion creep with n ≈ I is a fairly common occurrence. For example, diffusion creep has been reported in Al_2O_3, BeO, Cr_2O_3, MgO, ix SiC, Si_3N_4, ThO_2, UO_2, and ZrO_2 stabilised with 25 mol.-% Y_2O_3. It should be noted that Harper- Dorn creep leads also to a stress exponent of n = I, but there are very few experimental reports of this type of creep in ceramics and it will not be considered further.

One of the major difficulties in comparing experimental results with diffusion creep theories is the paucity of data on grain boundary diffusion coefficients, especially since it was noted recently that it is not generally possible to obtain meaningful information on diffusion coefficients from experimental creep da ta. As a result, a meaningful comparison is possible only for conditions of Nabarro- Herring creep with n ≈ l and p ≈ 2.

In Figure, experimental data for alumina doped with MgO are shown plotted in the logarithmic form of normalised creep rate ($\dot{\varepsilon}$ kT/DGb)(d/b)² versus normalised stress σ/G. The theoretical prediction for equation $\dot{\varepsilon} = 9.3 \dfrac{D_1 Gb}{kT} \left(\dfrac{b}{d}\right)^2 \left(\dfrac{\sigma}{G}\right)$ is plotted as a broken line marked Nabarro- Herring.

It can be seen from inspection of Figure that, despite scatter in the data points, the experimental results are in agreement with the theoretical predictions to within two orders of magnitude.

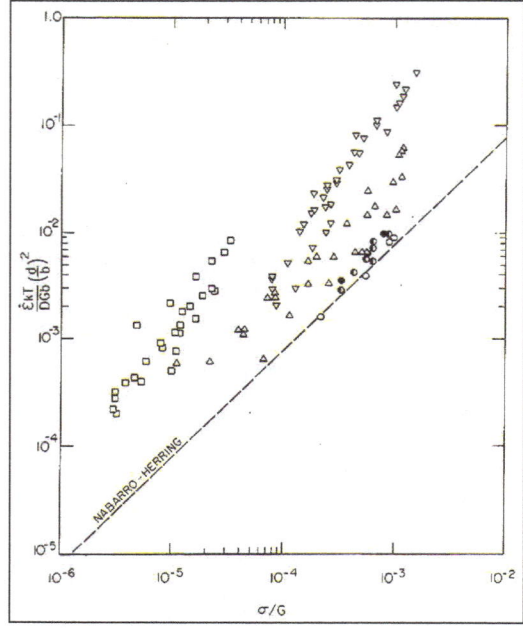

Normalised strain rate versus normalised stress for Al_2O_3 doped with MgO showing n $\approx$ 1.

	T, K	d, μm	MgO, ppm
○	1873	18-19	400
◐	1923	14-30	400
□	1973	18	400
△	1873-2073	7-34	...
▽	1573-1823	1.2-1.3	2500
◻	1723-2023	15	1000

Two factors contribute to the apparent scatter in Figure. First, several sets of experimental results have demonstrated that, at any selected stress level, the measured creep rates in tension are faster than in compression. A similar tendency can be seen in Figure, where the rates in tension are faster than in bending and the latter are faster than in compression. It should be noted that, as reported experimentally in bending and tensile tests on alumina, each set of data gives a similar value for n. Second, it is known that the diffusion coefficients are sensitive to the level of impurities but it was not possible to include this variation in the plotting of Figure.

Finally, it is important to note that apparent scatter may arise in the creep data for materials such as UO_2 because of small deviations from stoichiometry and the consequent effect on the values of the diffusion coefficient.

The preceding sections show that ceramics often exhibit stress exponents of 5, 3, or 1: the precise significance of these different types of behaviour is now examined.

Significance of n ≈ 5

In metals, a stress exponent close to 5 is associated with the formation of a subgrain structure in which the grains become divided into subgrains with the subgrain boundaries having low angles of misorientation (typically < 2°). It was shown for metals, and confirmed for ceramics, that the average subgrain size $i.$ is related inversely to the stress by an expression of the form,

$$\frac{\lambda}{b} = \xi\left(\frac{\sigma}{G}\right)^{-1}$$

where, the constant ξ has a value of ~20.

Observations on metals show also that the normalised density of dislocations within the subgrains $b\rho^{1/2}$ may be expressed as,

$$b\rho^{1/2} = \psi\left(\frac{\sigma}{G}\right)$$

where, ρ is the dislocation density and ψ is a constant having a value close to unity. Again, a similar relationship is applicable in ceramics.

The close similarity in the stress exponents and substructural details of the creep behaviour of metals and ceramics when $n \approx 5$, combined with the similar normalised creep rates in ceramics and nominally pure fee metals, provides strong support for a similarity in the rate controlling deformation mechanism. It is therefore concluded that ceramics with $n \approx 5$ deform by the glide and climb of dislocations controlled by climb.

Significance of n ≈ 3

Many metallic solid solution alloys exhibit $n \approx 3$ due to a process in which solute drag limits the rate of dislocation glide. However, this process is not likely to occur in ceramics because there is often little or no solute and, unlike metals, there is no experimental evidence for a transition from $n \approx 3$ to $n \approx 5$ with a decrease in stress as expected from the sequential nature of the glide and climb processes. Furthermore, there is experimental evidence for subgrain formation in MgO when $n \sim 3$, although subgrains are not formed in metallic alloys when $n \approx 3$.

It has been noted that $n \approx 3$ is generally associated with ceramics in which the anion/cation ratio r_a/r_c is > 2 whereas $n \approx 5$ occurs in materials having $r_a/r_c < 2$. The correlation between the r_a/r_c ratio and the value of n is related to the slip character of ceramics, leading to the more general observation that $n \approx 5$ is associated with the presence of five independent slip systems whereas $n \approx 3$ is associated with either a lack of five independent systems or, if five independent systems are available, to a lack of interpenetration of these systems. A polycrystalline material must have five independent slip systems in order to satisfy the von Misescriterion for homogeneous deformation. Thus, a stress exponent of $n \approx 5$ is associated with fully ductile behaviour whereas $n \approx 3$ leads to creep deformation which is for a less ductile condition.

The validity of this conclusion may be demonstrated by two examples. First, MgO deforms at low temperatures on the {110} $<1\bar{1}0>$ slip system which gives two independent slip systems. At T > 2000 K, slip is activated on both the {001} (110) and {110} $<1\bar{1}0>$ systems giving five independent systems. This is in agreement with the observation that the brittle- ductile transition in MgO is ~2000 K, and it is consistent also with the reports of a stress exponent of n ≈ 3 in MgO at typical creep testing temperatures of < 1800 K.

Second, UO_2 deforms by slip at high temperatures on the {001} $<1\bar{1}0>$, {110} $<1\bar{1}0>$, and {111} $<1\bar{1}0>$ systems, thereby giving the five independent slip systems necessary for homogeneous deformation. The brittle- ductile transition occurs in UO_2 at ~ 1420 K. which is consistent with the observation of n ≈ 5 at testing temperatures above ~1500 K.

The very good agreement between the theoretical predictions and the experimental results showing n ≈ 3, together with the consistency in terms of the available slip systems, leads to the conclusion that creep with n ≈ 3 occurs by the climb of dislocations from Bardeen- Herring sources under experimental conditions where the materials do not have five independent and interpenetrating slip systems. It has been demonstrated that general deformation can be accomplished in polycrystalline materials with less than five independent slip systems by dislocation climb from Bardeen- Herring sources.

Significance of n ≈ 1 and Role of Interface Reaction Control

Experimental results showing n ≈ 1 are usually attributed to some form of diffusion creep. Many experimental results on ceramic materials give n ≈ 1 and values of p of about 2–3 and this is consistent with the theories of diffusion creep. However, it is important to note that, as can be seen in Table, Harper- Dorn creep leads also to n ≈ 1, although deformation then occurs by an intragranular dislocation process. Harper- Dorn creep can be distinguished from diffusion creep because there is no dependence on grain size and p = 0.

There is also a consistent body of creep data giving values of n close to 2. Some of these results are complicated by the presence of at least a partial grain boundary glassy phase but other results occur in materials where there is no glassy phase.

In the absence of a glassy phase, the results with n ≈ 2 may be interpreted in terms of diffusion creep with control by an interface reaction. If this reaction involves the formation or annihilation of vacancies, it occurs sequentially with the diffusion creep process and therefore it leads to creep rates that are slower than those predicted by the theories for diffusion creep.

Unfortunately, it is usually difficult to check whether the rates are slower than the theoretical predictions because of uncertainties in the values of the diffusion coefficients. An alternative verification may be achieved by noting that there should be a transition with increasing stress from interface reaction controlled creep with n ≈ 2 to diffusion creep with n ≈ 1. An example of this type of transition is available for alumina.

Transitions Increep Behaviour

The preceding discussion summaries the rate controlling mechanisms for the separate conditions where the values of n are close to ~5, ~3, and ~1, respectively. In practice, however, there is an anticipated transition between diffusion creep and power law creep as illustrated schematically in Figure.

In ceramics having five independent lip systems which interpenetrate, power law creep occurs at high stresses with $n \approx 5$ due to the glide and climb of dislocations controlled by the climb process. This is illustrated in Figure. As the stress level is reduced, there is a transition to diffusion creep with $n = 1$. At even lower stresses, there is the possibility of an increase in the stress exponent either to a value of - 2 corresponding to interface reaction control or to a value which increases towards infinity due to the presence of a threshold stress.

The precise appearance of the curve depends also on the grain size because diffusion creep has $p = 2$ or 3 for Nabarro- Herring and Coble creep, respectively, whereas power law creep has $p = 0$. Therefore, it is experimentally easier to reveal diffusion creep when the grain size is small and power law creep when the grain size is large.

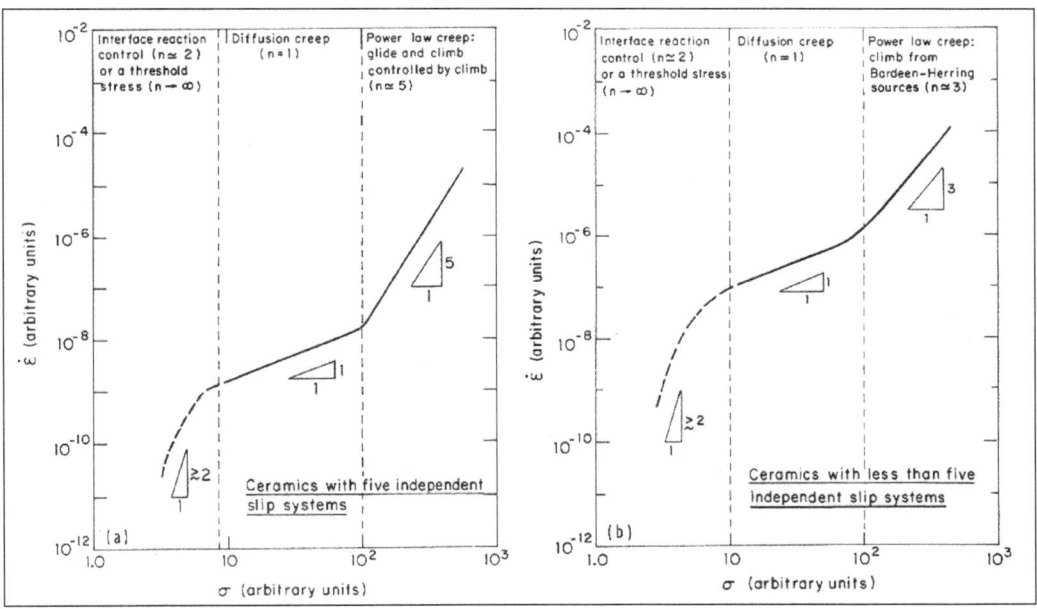

Schematic illustration of strain rate versus stress for ceramics with a five independent slip systems and bless than five independent slip systems.

In ceramics having less than five independent slip systems, or if five independent slip systems do not interpenetrate the situation is essentially similar except that power law creep occurs through the climb of dislocations from Bardeen-Herring sources. This condition is illustrated in Figure.

Creep experiments are often performed over a sufficiently restricted range of stress that the transitions shown in Figure. are not revealed. However, Figure gives an example of a transition from $n \approx 5$ to $n \approx 1$ with decreasing stress for UO_2 tested at a temperature of 1808 K. By linearly replotting the data points in the regime of $n \approx 1$ as strain rate against stress, as shown in Figure, it is apparent that there is no evidence for a threshold stress either in the experiments on UO_2 where $n \approx 5$ at high stresses or in experiments on Al_2O_3 where Figure indicates that $n \approx 3$.

When the grain size is increased. For the experiments on UO_2, a detailed analysis of data obtained at different temperatures suggests that Coble creep makes a significant contribution. It is possible to predict the magnitudes of the transition stresses from diffusion creep to power law creep depicted in Figure.

When there are five independent slip systems, equation $\dot{\varepsilon} = 9.3 \dfrac{D_{com} Gb}{kT} \left(\dfrac{b}{d}\right)^2 \left(\dfrac{\sigma}{G}\right)$ gives the rate of diffusion creep and equation $\dot{\varepsilon} = A \dfrac{DGb}{kT} \left(\dfrac{b}{d}\right)^p \left(\dfrac{\sigma}{G}\right)^n$ can be used for power law creep with A= A_{p_1}, p = 0, n = 5 and assuming D = $D_{a(I)}$ for lattice diffusion of the (generally) slower moving anion. By equating these two relationships, and assuming diffusion creep is controlled by cation diffusion through the lattice, the normalised stress associated with the transition in Figure is given by:

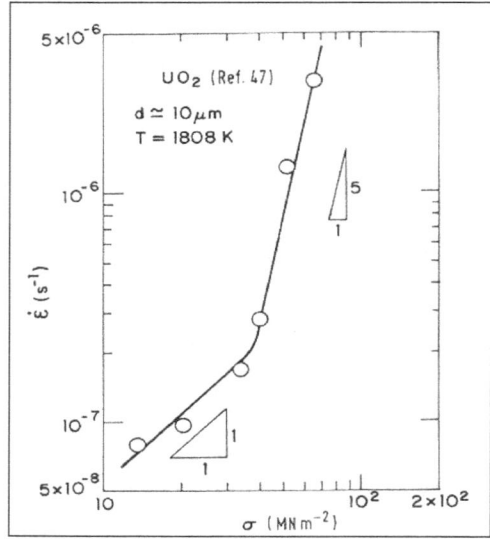

Strain rate versus stress for UO$_2$ showing transition from n $\approx$ 5 to n $\approx$ 1 with decreasing stress.

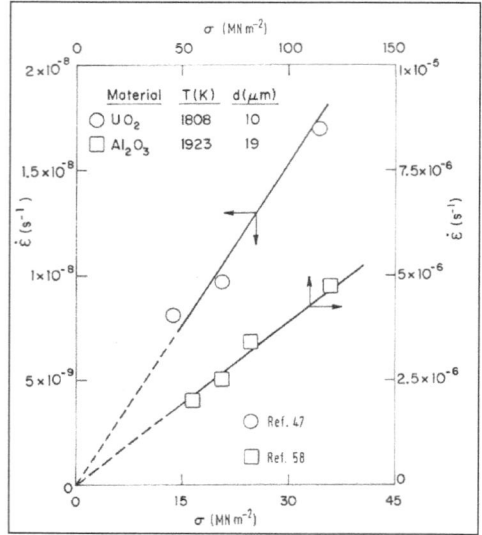

Strain rate versus stress for Uo$_2$ and Al$_2$o$_3$ showing lack of threshold stress.

$$\left(\frac{\sigma}{G}\right)_t = \left[\frac{9.3}{Ap_1}\left(\frac{\beta}{\propto}\right)\left(\frac{b}{a}\right)^2 \frac{D_{e(1)}}{D_{a(1)}}\right]^{0.25}$$

Alternatively, when there is less than five independent slip systems, equations $\dot{\varepsilon} = 0.22 \dfrac{D_1 Gb}{kT}\left(\dfrac{\sigma}{G}\right)^3$ and $\dot{\varepsilon} = 9.3 \dfrac{D_{com} Gb}{kT}\left(\dfrac{b}{d}\right)^2\left(\dfrac{\sigma}{G}\right)$ can be used under the same assumptions to give the normalised transition stress shown in figure.

$$\left(\frac{\sigma}{G}\right)_t = 6.5\left(\frac{b}{d}\right)\left(\frac{D_{c(1)}}{D_{a(1)}}\right)^{0.5}$$

Finally, it should be noted that diffusion creep is a common observation in laboratory experiments on ceramics whereas it is seldom reported for metals. This is partly due to the smaller grain sizes in many ceramics and the consequent increase in the transition stresses as indicated by inspection of equations $\left(\dfrac{\sigma}{G}\right)_t = \left[\dfrac{9.3}{Ap_1}\left(\dfrac{\beta}{\propto}\right)\left(\dfrac{b}{a}\right)^2\dfrac{D_{e(1)}}{D_{a(1)}}\right]^{0.25}$ and $\left(\dfrac{\sigma}{G}\right)_t = 6.5\left(\dfrac{b}{d}\right)\left(\dfrac{D_{c(1)}}{D_{a(1)}}\right)^{0.5}$. It is also partly due to the possibility in ceramics of fast diffusion of one of the ionic species along the grain boundaries. Since diffusion creep is usually dominated by lattice diffusion of the cation and power law creep with $n \approx 5$ is usually dominated by lattice diffusion of the anion, it follows that, taking a typical value of $D_{c(1)}/D_{a(1)} \approx 10^2$, the value of $(\sigma/G)_t$ in equation $\left(\dfrac{\sigma}{G}\right)_t = \left[\dfrac{9.3}{Ap_1}\left(\dfrac{\beta}{\propto}\right)\left(\dfrac{b}{a}\right)^2\dfrac{D_{e(1)}}{D_{a(1)}}\right]^{0.25}$ is increased by a factor of ~ 3. By contrast, equation $\left(\dfrac{\sigma}{G}\right)_t = \left[\dfrac{9.3}{Ap_1}\left(\dfrac{\beta}{\propto}\right)\left(\dfrac{b}{a}\right)^2\dfrac{D_{e(1)}}{D_{a(1)}}\right]^{0.25}$ for metals reduces to,

$$\left(\frac{\sigma}{G}\right)_t\left[\frac{9.3}{A_{p1}}\left(\frac{b}{d}\right)^2\right]^{0.25}$$

Which is independent of the diffusion coefficient because Nabarro- Herring diffusion creep and power law creep are both dependent upon the lattice diffusivity.

Fracture of Ceramics

Fracture is the most important aspect of the mechanical properties of structural ceramics.

Ductile and Brittle Fracture

The physical separation of a component into two or more pieces is termed as fracture. In general, if a metal specimen is deformed in a tensile machine, it undergoes substantial plastic deformation as manifested by reduction in its cross sectional area and elongation of its gage length. Such a fracture is called a ductile fracture. In other materials, such as glass, there is no significant plastic deformation before fracture. Such a fracture, accompanied by little or no plastic deformation, is called a brittle fracture. In addition to glass, which is an amorphous material, the brittle fracture can also take place in crystalline materials such as ionic solids, ceramics, bcc and hcp metals at low temperatures and polymers below their glass transition temperature.

Theoretical Fracture Strength

Assuming that the brittle fracture occurs by the rupture of atomic bonds, an estimate of the stress required for brittle fracture can be obtained. Figure shows schematically the atomic bonds in a crystalline solid being stretched due to an applied force. In figure, the situation for a single bond is considered. Figure, upper figure, shows the change in the potential energy φ as the distance between the atoms changes while the lower figure shows stress as a function of separation between the atoms. The stress can be approximated by a sin function as,

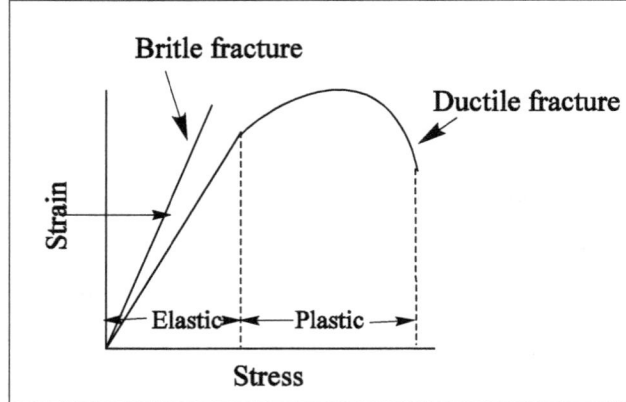

Tensile test plots for a ductile and a brittle material. The brittle material fails without any plastic deformation.

$$\sigma = \sigma_{max} \sin \frac{\pi(a - a_o)}{a_m}$$

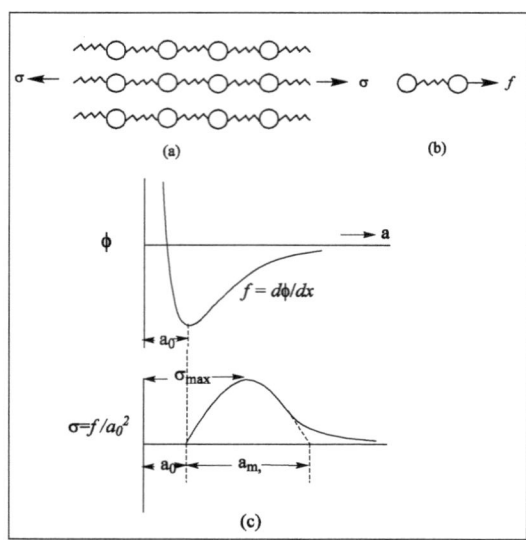

(a) A crystalline solid is stresses; the atomic bonds are represented as stretched springs (b) a single bond (c) variation in the potential energy of the atoms in (b) with their separation (upper figure) and variation in the stress as a function of the atomic separation.

Here σ_{max} is the maximum stress reached during bond stretching before the bonds break i.e. it is the fracture stress, a_o is the interatomic separation, am is as shown in the figure. When the bonds break a fracture surface is produced. The work done in bond breaking is the energy of the new surface created. As two fracture surfaces are created we have,

$$2\gamma = \sigma_{max} \int_{a_o}^{a_o+a_m} \sin\frac{\pi(a-a_o)}{a_m} da = 2\frac{a_m}{\pi}\sigma_{max}$$

$$a_m = \pi\gamma / \sigma_{max}$$

for small strain, $\varepsilon = (a - a_o)/a_o$,

and $\sigma = \sigma_{max}\pi(a-a_o)/a_m == \sigma_{max}\pi a_o\varepsilon / a_m$,

which is of the form $\sigma = E\varepsilon$, where E is the Young's modulus.

Thus $E = \sigma_{max}(\pi a_o / a_m) = \sigma_{max}(\pi a_o / \pi\gamma)\sigma_{max}$

Or,

$$\sigma_{max} = \sqrt{\frac{E\gamma}{a_o}}$$

Using typical values of γ, σ_{max} = E/10. For the polycrystalline ceramics, E = 100 to 500 GPa. So the theoretical fracture stress, σ_{max}, should be between 10 to 50 GPa. Usually it is 100 to 500 MPa, i.e. less by a factor of 100.

Preexisting Cracks in Brittle Solids

Besides the large discrepancy between the theoretical and experiment strengths, there are other aspects of brittle fracture which need to be explained. These are exemplified by the following observations for silica glass:

- The strength of brittle solids has a large variation. Thus the strength of silica glass off the shelf varies from about 0.05 GPa to 0.1 GPa. There is a variation is strength of a given set of samples by a factor or 2 or more.

- The strength is smaller for a large sample size. Thus if a glass rod is broken in two, the strength of the resulting pieces is found to be larger than the parent piece. If this piece is broken yet again, the strength of the resulting pieces is found to be again higher.

The explanation for these observations was first provided by Griffith. During his work on the strength of glass, Griffith noticed that the strength depended on the condition of the surface of the glass. Thus a freshly drawn glass fibre had a very large strength, approaching the theoretical strength but even a little handling resulted in a considerable lowering of the strength. Griffith proposed that there are pre-existing cracks on the surface of the glass. He further proposed that on loading, a pre-existing crack would propagate and cause fracture if the decrease in the elastic strain energy due to its propagation is at least equal to the energy required to create the new crack surface as a result of fracture.

The glass rod in the above example has a distribution of surface cracks. On loading, the most severe of these cracks would propagate and cause fracture. When the sample is broken in two, and one of the pieces is retested, the strength is higher because the crack which now causes failure is less severe than the crack which caused the fracture of the parent piece.

While in a sample of glass, the pre-existing cracks are nearly all on the surface, in case of ceramics, there can be flaws both on the surface as well as in the interior of the ceramic. Ceramics are prepared by a powder processing method. The process inherently leads to presence of some pores and other flaws. The sintering at high temperatures during the preparation of a ceramic body causes the generation of grooves at the grain boundaries due to preferential sublimation from the grain boundaries. If the piece is subjected to grinding or machining, then also some surface flaws are introduced. The length of these flaws scales as the grain size. On loading, it is the most sever flaw which propagates and causes fracture.

Quantitative Treatment of Brittle Fracture

The fact that the fracture in the brittle solids initiates from the most severe flaw and that there is a distribution of the severity of these flaws in a given piece and further that this distribution varies from piece to piece raises the question whether there is any inherent property of a material which can be measured and would give an idea of its behaviour in brittle fracture. There are several approaches to the quantitative treatment of brittle fracture. Out of these the concept of stress intensity factor or the fracture toughness has proven to be most useful for the design purpose.

Treatment of the Griffith Crack: The Energy Balance Approach

Consider a plate in which there is no crack which is loaded to a uniform stress σ. The elastic energy per unit volume in the plate is then $\sigma^2/2E$ where E is the Young's modulus. Now introduce a "through the thickness" crack of length 2a in the center of the plate. The stresses in the plate in the neighbourhood of the crack relax because of the presence of the crack.

The relaxed zone on both sides of the crack can be approximated by a triangular shape of height βa i.e. in this zone the stress is zero. Then the relaxed volume to the right of the y axis = $\beta a^2 t$, where t is the thickness of the plate.

Reduction in the stored elastic strain energy per tip = $(\sigma^2/2E)\beta a^2 t$. Energy released (reduced due to the introduction of the crack) per unit thickness per tip = $(\sigma^2/2E)\beta a^2$

Accurate relation by Griffith gave this quantity to be = $(\sigma^2/2E)\pi a^2 = U$. There is an increase in energy due to the creation of the new crack surface per tip which is = $2\gamma a$. Total energy change is given by,

$$T = 2\gamma a - \left(\sigma^2/2E\right)\pi a^2$$

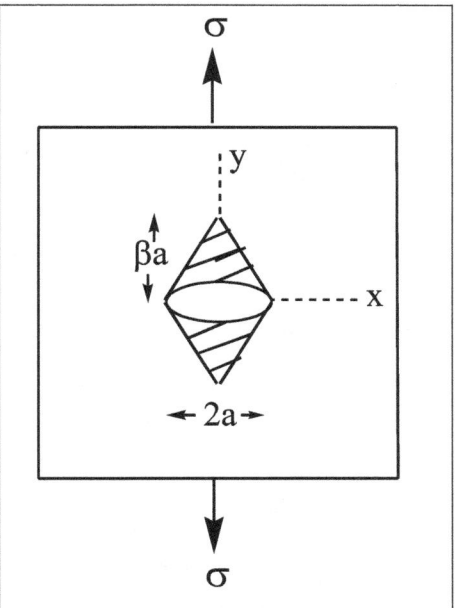

A plate is stressed uniformly. Introduction of a through the thickness elliptical crack in the centre of the plate produces a zone of reduced stress. It is approximated by a stress free triangular zone on both sides of the crack as shown by the shaded area.

Change in the total energy as the crack length increases by an incremental amount da is,

$$dT / da = 2\gamma - \left(\sigma^2 / E\right)\pi a$$

The change in the total energy T with increasing crack length is shown in Figure. After the crack length exceeds a critical value a^*, there is decrease in energy with further increase in the crack length, i.e. the crack propagates and the fracture occurs. At this point ,

$$dT / da = 0 = 2\gamma - \left(\sigma^2 / E\right)\pi a$$

Or

$$2\lambda = \left(\sigma^2 / E\right)\pi a$$

The right hand side of this equation is called the crack extension force and is denoted by G,

$$G = \left(\sigma^2 / E\right)\pi a$$

In reality, in addition to the creation of the new crack surface, additional events occur which need energy as the crack advances. In metals there is plastic deformation ahead of the crack tip. In glass and ceramics, there may be generation of microcracks and other events. To take this into account, 2γ is replaced by a general term R, called the crack extension force. Then R is the energy which is required for the crack to advance; for the crack to advance G must be equal to or more than R required for the crack to advance. Thus for the failure to occur, G must exceed R.

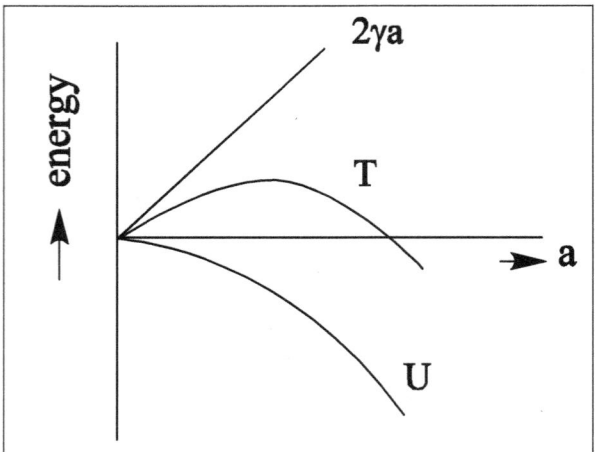

Changes in the surface energy (2γa), stored elastic strain energy (U)
and the total energy as the crack advances.

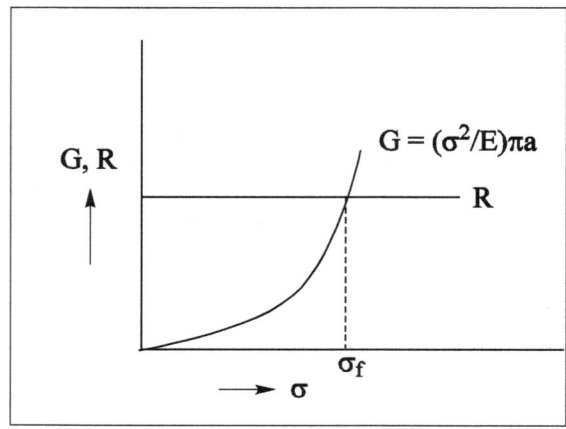

R and G vs. applied stress, σ.

The situation when the stress is gradually increased with a crack of certain length preexisting in the body is shown in Figure which shows schematically the change in G with applied stress. At a stress equal to σ_f, G exceeds R and the fracture occurs. At this point,

$$G_c = \left(\sigma_f^2 / E\right)\pi a$$

We can also write,

$$\sqrt{G_c E} = \sigma_f \sqrt{\pi a}$$

The term on the left hand side is a function of the properties of the material while that on the right hand side depends on external variables. This approach provides useful insight into the fracture process. However, the calculation of the energy term in different geometries is not straightforward.

Stress Intensity Factor Approach

This approach has turned out to be most useful as it provides a directly measurable quantity which can be used for the design purposes. In this approach a mathematical, infinitely sharp crack of length 2c is considered. The stresses near the tip of such a crack can be calculated by the elasticity theory. The stresses on element located at a point P near the crack tip are given by the following expressions:

$$\sigma_x = \sigma \left(\frac{c}{2r} \right)^{1/2} \left[\left(1 - \sin\frac{\vartheta}{2}\sin\frac{3\vartheta}{2} \right) \right]$$

$$\sigma_y = \sigma \left(\frac{c}{2r} \right)^{1/2} \left[1 + \sin\frac{\vartheta}{2}\sin\frac{3\vartheta}{2} \right]$$

$$\sigma_{xy} = \sigma \left(\frac{c}{2r} \right)^{1/2} \sin\frac{\vartheta}{2}\cos\frac{\vartheta}{2}\cos\frac{3\vartheta}{2}$$

$\sigma_z = 0$ for plane stress; $\sigma_z = v(\sigma_x + \sigma_y)$ for plane strain.

In the first three expressions, the quantity $\sigma\sqrt{c}$ is common in all expressions. It is convenient to define a quantity $\sigma\sqrt{\pi c}$ as K_I. This quantity is called the mode I stress intensity factor. Mode I refers to the case when the stress is applied perpendicular to the crack surface. In Mode II and Mode III, the stress is applied parallel to the crack surface producing a shear loading and a tear loading respectively. We will confine ourselves in this discussion to Mode I only.

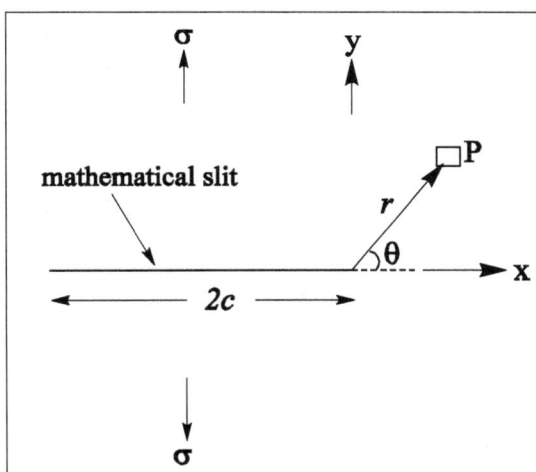

A body with a mathematical slit of length 2c is remotely loaded with a stress σ. Stresses near the tip on an element at point P can be calculated.

In terms of K_I the various expressions for stresses near the crack tip assume the form,

$$\sigma_x = \frac{K_I}{\sqrt{2\pi r}} \left[f(\vartheta) \right]$$

Earlier we saw that the critical energy release rate for fracture is,

$$G_c = \left(\sigma_c^2 / E \right) \pi a$$

This can be written in terms of K_I as,

$$Gc = \left(\sigma_c^2 / E\right)\pi a = K_{Ic}^2 / E$$

This implies that the fracture would occur when the K_I reaches a critical value, K_{IC}, since then the energy release rate would reach the critical value for fracture. The quantity K_{IC} is called the critical stress intensity factor or the fracture toughness T of the material.

Design with Preexisting Cracks

For real situations, the stress intensity factor depends on the geometry of the cracked configuration. For a body with a through the thickness crack of length 2a, the stress intensify factor is given as,

$$K_I = Y\sigma\sqrt{\pi a}$$

Here Y is a factor which depends on the dimensions of the sample and the loading geometry.

Example: A sample of a brittle solid has four cracks of sizes 0.5 μm, 0.9 μm, 1.7 μm and 2.4 μm, all loaded in mode I. The fracture toughness, K_{IC} of the material is 12 MPam1/2. What should be the strength of the sample? Assume the geometrical factor to be 1.1.

Solution: It is the largest crack which results in failure. In this case 2a = 2.4 μm or a = 1.2 μm.

Substituting in above example we get,

$$12\times10^6 = 1.1\ \sigma_f \left(\pi\times1.2\times10^{-6}\right)^{1/2}$$

Or

$$\sigma_f = 5619 \text{ MPa}.$$

In real situations, it may be possible to determine the distribution of crack lengths by non-destructive testing. Then knowing the value of the K_{IC} for the material, the largest stress that can be tolerated by the structure can be determined.

The R Curve Behaviour

Previously, we defined the quantity R, the crack growth resistance as the energy which is required for the crack to advance i.e. it is the rate (per unit increase in crack length) of release of energy required for the crack to advance. In Figure, R is shown as a horizontal line, i.e. it is independent of the crack length. For a body containing a crack of half-length a_1 and subjected to a stress σ_1, the quantity G = ($\sigma_1\ 2\pi a_1/E$) is just equal to R at a = a_1. The crack here becomes unstable to a slight increase in its length and so grows in an unstable manner as G remains more than R and the fracture occurs. If the crack length were a_2 initially, then a stress σ_2 is required for the fracture to occur. At a lower stress, R > G and the fracture does not occur.

In the above treatment, R is assumed to be independent of the crack length. This is the case, for example, when there is no plastic deformation or any other energy dissipating process occurring ahead of the crack tip. Then R is equal to 2γ, where γ is the surface energy of the solid and is independent of a, the crack length. However, in almost all cases, there are energy dissipating processes occurring near the crack tip as the crack advances. In case of metals, there is a zone at the crack tip in which plastic deformation occurs as the crack advances. In ceramics also there are processes such as micro-cracks formation, stress induced phase transformation, etc. which dissipate energy. Thus the energy required for the crack to advance is much larger than 2γ. The crack growth resistance is then the sum of the surface energy term and all the other energy terms due to these crack tip processes.

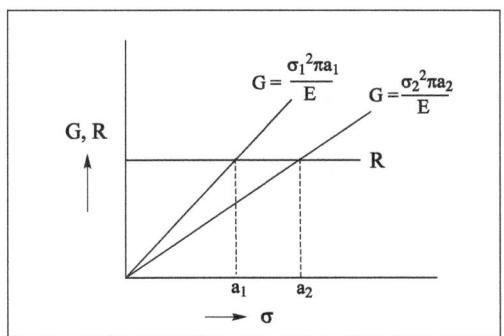

Energy release rate G and crack growth resistance R vs. crack length a. Cracks of half lengths a_1 and a_2 will extend unstably when the stress is raised to σ_1 and σ_2 respectively.

The value of R may still be independent of a if the volume of the zone ahead of the crack tip in which the energy dissipating processes occur remains constant as the crack advances; on the other hand, if it increases with increasing crack length, then R will also show an increase with the crack length. In this case, the crack growth occurs in a stable manner i.e. on increasing the stress to a certain value, the crack grows a little and then stops unless the stress is further raised. When the crack is extending, $R \le G$. If the crack stops, it means that R has now become larger than G i.e. R increases as the crack grows. Since and as σ and a are both increasing, G increases nonlinearly with a, hence R also increases nonlinearly with a as shown in figure below.

$$G = \pi\sigma^2 a / E$$

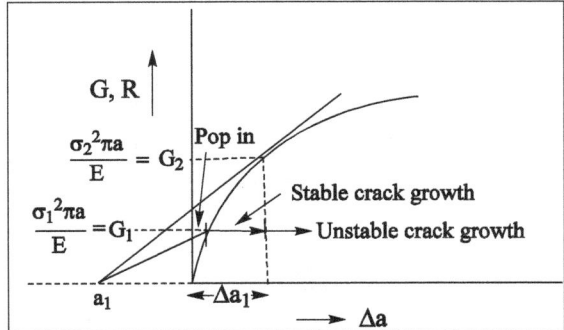

Increase in crack length. Δa, vs. G, R for nonlinear R curve.

It is more illustrative to plot Δa, the increase in the crack length, rather than a, on the x axis; furthermore, since the 2γ term is generally negligible in comparison to the energy dissipated by the crack tip processes, R can be taken to start from the origin at $\Delta a = 0$ as shown in Figure. The initial crack lengths are indicated on the left side on the x axis. Consider a crack of initial half-length a_1. When the sample is loaded to a stress say, σ_1, the value of G immediately reaches

G1 as shown by the intersection of the line drawn from a_1 with the R curve and the crack length increases by a small amount Δa_1. This is said to be the initial "pop-in". The line, if extended, shows that G now becomes less than R and so the crack stops. If the stress is now increased, the crack extends until a value of $G = G_2$ is reached. This value is given by the tangent drawn to the R curve from the point a_1. As, on extending, this line stays above the R curve, it implies that as the crack extends at this stress, the G is always more than R and so the crack extends unstably i.e. the fracture occurs at this stress.

Fracture Toughness Testing

Various geometries of the test sample have been designed for the determination of the critical stress intensity factor, K_{IC}. This quantity is also called the fracture toughness. A very simple and common geometry is that of a bar of the material with a rectangular cross section tested in three point bending. The sample dimensions and the loading geometry is shown in Figure.

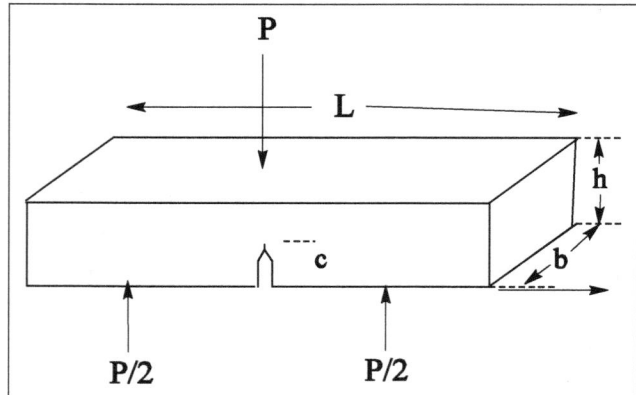

Loading geometry in three point bending. The load is applied through a roller on the top surface. The beam is supported on two rollers.

For this case, the geometric factor is given by a polynomial expression. For the case when L/h = 10, this expression is,

$$Y = 1.93 - 3.07\left(\frac{c}{h}\right) + 14.53\left(\frac{c}{h}\right)^{2-} 25.11\left(\frac{c}{h}\right)^{3} + 25.8\left(\frac{c}{h}\right)^{4}$$

In addition to three points bending, several other geometries are used for the fracture toughness testing of ceramics.

Another popular method for measuring the fracture toughness of ceramics is by indentation. This technique evolved from the need to understand the fracture behaviour due to surface cracks as it was noted that the fracture in ceramics usually occurred from surface cracks which formed due to contact damage. To simulate such cracks an indentation technique was used in which the surface of a specimen is indented with a hardness testing indenter, usually the Vickers indenter which has the shape of a pyramid with a square base.

When the surface of a ceramic is indented by such an indenter, cracks are formed at right angles to the surface which are semicircular in shape; these are called median radial crack. In addition to these cracks, lateral cracks which are parallel to the surface are also formed, which may cause spalling of the surface; these are not of interest in fracture toughness testing.

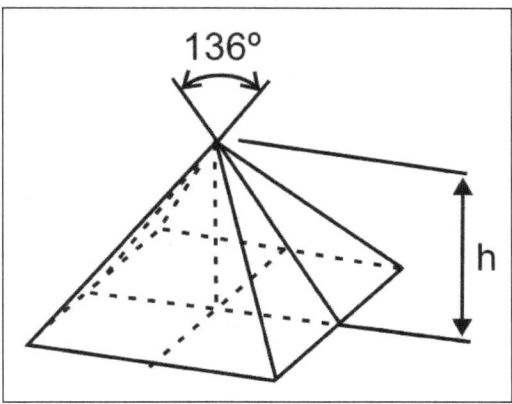

A Vickers's indenter.

The impression of the indent and the situation below the indentation impression is shown in Figure. Cracks can be seen emanating from the corners of the square indentation. The length of these cracks is shown to be equal to be 2c in the figure. In actual situations, they may be slightly different in length; in that case an average of the two can be taken to calculate the fracture toughness.

During indentation, the region just next to the indent is deformed plastically. Next to this plastically deformed region is an elastically deformed region. When the indentation load is released, the elastically deformed region tries to regain a zero strain position but is unable to do so due to the plastically deformed region. The plastically deformed region acts as a wedge and opens up the radial median cracks. A residual stress is also produced as shown in Figure.

The stress intensity factor due to the residual stress acting on a crack of length 2c is given by,

$$K_I = \frac{\chi P}{c^{3/2}}$$

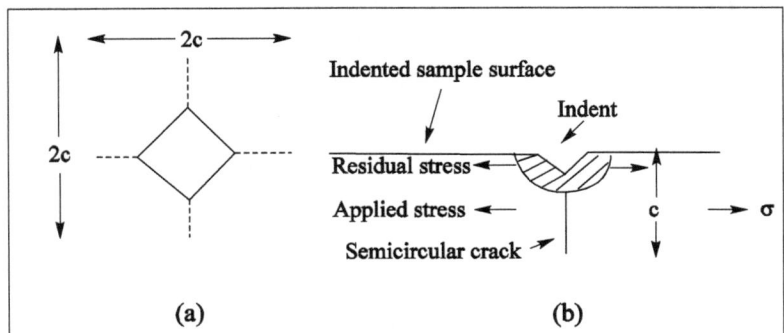

(a) Appearance of the indent on the indented surface; the lines emanating from the corners are the radial median semicircular cracks (b) region below the indent when an external load is also applied during the testing of the sample subsequent to indentation.

Here P is the applied load during indentation and χ is a parameter which depends on the material and the indenter. When the indentation load is released, initially the crack length is small and KI is large, larger than R, the crack growth resistance. The crack therefore grows in length which causes a decrease in K_I according to the above equation. The crack growth stops when K_I drops and becomes equal to R, the crack growth resistance. This also equal to the fracture toughness T. The fracture toughness is then given by the equation $K_I = \dfrac{\chi P}{c^{3/2}}$.

The parameter χ is given by,

$$\chi = \beta \sqrt{E/H}$$

Here E is the Young's modulus and H is the hardness of the material. β is a parameter which is determined from calibration tests and is usually equal to 0.016. The fracture toughness can also be determined by loading the indented specimen with an initial crack of length $2c_0$ and determining the stress for failure, σ_f. The fracture toughness T in this case is given by,

$$\sigma_f = \frac{0.47 T^{4/3}}{Y(\chi P)^{1/3}}$$

It can also be shown that $c_m \sim 2.5\, c_0$ where $2c_m$ is the crack length at fracture.

Statistics of Brittle Fracture

A brittle solid contains a large number of cracks of varying severity. Upon loading, the most severe of these cracks causes fracture. The distribution of crack lengths varies from sample to sample. Figure shows schematically how the crack length distribution may appear. As the length of the largest crack present in a group of samples from the same ceramic may vary widely, their strength will also have a distribution. In the brittle materials the scatter in the fracture strength is quite wide, unlike the case for, say, metal specimens where the scatter in the strength is very little. A simple average of the strengths in case of a brittle solid is not of much use. We need more quantitative information about the distribution of strengths and about the reliability of the material.

Since the strength depends on the extreme value of the crack length in a sample, it implies that the strength of a group of otherwise similar samples will have a distribution similar to the distribution of the extreme value of the crack lengths in the group. One of the statistical distributions which are found to be followed by the extreme value of the crack lengths in the brittle solids is the Weibull distribution.

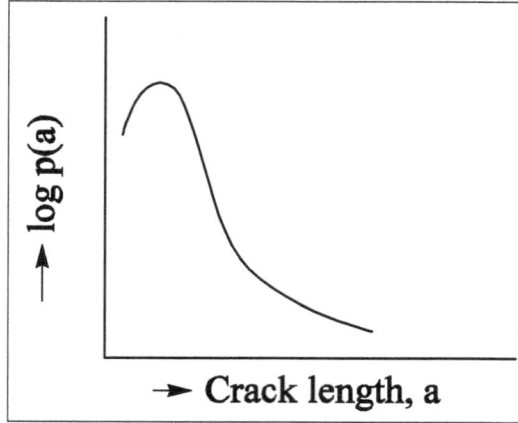

Crack length vs. log of the probability of a crack of that length being present in the sample.

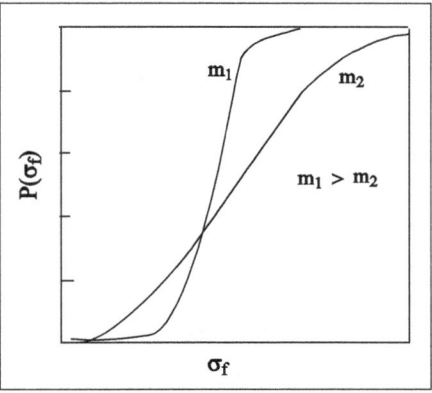

Fracture strength data for two materials fitted to Weibull distribution; the material with the shape parameter m_1 has a narrow distribution of fracture strengths and is more reliable.

When the samples are all of the same volume, then the following simple, two parameter form of the Weibull distribution can be used:

$$P(\sigma_f) = 1 - \exp\left[-\left(\frac{\sigma_f}{\sigma_o}\right)^m\right]$$

Here $P(\sigma_f)$ is the probability that the sample fails at a stress $\leq \sigma_f$ and σ_o and m are the parameters of the distribution. These parameters can be extracted from a data of the fracture stress values of a group of samples. For this, we take repeated log of the above equation and convert it into the form.

$$\ln[-\ln(1-P(\sigma_f))] = m \ln\sigma_f - m \ln\sigma_o$$

Then a plot of $\ln[-\ln(1-P(\sigma_f))]$ vs. $\ln \sigma_f$ is a straight line whose slope gives mad the intercept gives $-m \ln \sigma_o$.

The parameter m is called the shape parameter. Figure shows schematically the probability plots for two values of m. When m is large, the curve is steep and is in a narrow range implying that the strength values are distributed in a small range; when m is small, the curve spreads out and the scatter in the strength values is large. Thus m denotes the reliability of the material – a high value of m means a more reliable material with narrowly distributed values of the fracture strength. The value of m can be as high as 100 for metals while for ceramics it is usually 1 to 5. However, now careful fabrication can produce ceramics with a much higher values of m. The parameter σ_o is called the scaling parameter; it is the stress below which 63% of the samples fail.

References

- Electrical-and-Magnetic-Properties-of-Ceramics: inflibnet.ac.in, Retrieved 02, January 2020

- Characteristics-of-Creep-Deformation-in-Ceramics-233608856: researchgate.net, Retrieved 10, August 2020

3

Glass Ceramics

The polycrystalline materials which are produced through controlled crystallization of base glass are referred to as glass ceramics. A few of the topics studied under glass ceramics are kinetic theory of glass formation, structural theories of glass formation and applications of glass ceramics. This chapter closely examines these key concepts of glass ceramics to provide an extensive understanding of the subject.

Glass-ceramics are fine-grained polycrystalline materials formed when glasses of suitable compositions are heat treated and thus undergo controlled crystallisation to the lower energy, crystalline state. It must be emphasised here that only specific glass compositions are suitable precursors for glass-ceramics due to the fact that some glasses are too stable and difficult to crystallise whereas others result in undesirable microstructures by crystallising too readily in an uncontrollable manner. In addition, it must also be accentuated that in order for a suitable product to be attained, the heat-treatment is critical for the process and a range of generic heat treatment procedures are used which are meticulously developed and modified for a specific glass composition.

A glass-ceramic is formed by the heat treatment of glass which results in crystallisation. Crystallisation of glasses is attributed to thermodynamic drives for reducing the Gibbs' free energy, and the Amorphous Phase Separation (APS) which favours the crystallisation process by forming a nucleated phase easier than it would in the original glass. When a glass is melted, the liquid formed from the melting might spontaneously separate into two very viscous liquids or phases. By cooling the melt to a temperature below the glass transformation region it will result in the glass being phase separated and this is called liquid-liquid immiscibility. This occurs when both the phases are liquid. Hence a glass can simply be considered as a liquid which undergoes a demixing process when it cools. The immiscibility is either stable or metastable depending on whether the phase separation occurs above or below the liquidus temperature respectively. The metastable immiscibility is much more important and has two processes which then cause phase separation and hence crystallisation; nucleation and crystal growth and spinodal decomposition. The first APS process has two distinguished stages; Nucleation (whereby the crystals will grow to a detectable size on the nucleus) and Crystal growth. Nucleation can either be homogeneous; where the crystals form spontaneously within the melt or heterogeneous; crystals form at a pre-existing surface such as that due to an impurity, crucible wall etc. Many a time the parent glass composition is specifically chosen to contain species which enhance internal nucleation which in the majority of cases is required. Such species also called nucleating agents can include metallic agents such as Ag, Pt and Pd or non-metallic agents such as TiO_2, P_2O_5 and fluorides. The second process is spinodal decomposition which involves a gradual change in composition of the two phases until they reach the immiscibility boundary. As both the processes for APS are different, the glass formed will clearly result in having different morphology to each other.

A glass-ceramic is usually not fully crystalline; with the microstructure being 50-95 volume % crystalline with the remainder being residual glass. When the glass undergoes heat treatment, one or more crystalline phases may form. Both the compositions of the crystalline and residual glass are different to the parent glass. In order for glass-ceramics having desirable properties to be developed, it is crucial to control the crystallisation process so that an even distribution of crystals can be formed. This is done by controlling the nucleation and crystal growth rate. The nucleation rate and crystal growth rate is a function of temperature and are accurately measured experimentally.

The aim of the crystallisation process is to convert the glass into glass-ceramic which have properties superior to the parent glass. The glass-ceramic formed depends on efficient internal nucleation from controlled crystallisation which allows the development of fine, randomly oriented grains without voids, microcracks, or other porosity. This results in the glass-ceramic being much stronger, harder and more chemically stable than the parent glass. Glass-ceramics are characterised in terms of composition and microstructure as their properties depend on both of these. The ability of a glass to be formed as well as its degree of workability depends on the bulk composition which also determines the grouping of crystalline phases which consecutively govern the general physical and chemical characteristics, e.g. hardness, density, acid resistance, etc. Nucleating agents are used in order for internal nucleation to occur so that the glass-ceramic produced has desirable properties. Microstructure is the key to most mechanical and optical properties; it can promote or diminish the role of the key crystals in the glass-ceramic. The desirable properties obtained from glass-ceramics are crucial in order for them to have applications in the field of biomaterials.

Glass-ceramics are used as biomaterials in two different fields: First, they are used as highly durable materials in restorative dentistry and second, they are applied as bioactive materials for the replacement of hard tissue. Dental restorative materials are materials which restore the natural tooth structure (both in shape and function), exhibit durability in the oral environment, exhibit high strength and are wear resistance. In order for dental restorative materials to restore the natural tooth structure, it is crucial to maintain the vitality of the tooth. . However non-vital teeth may also be treated with restorative materials to reconstruct or preserve the aesthetic and functional properties of the tooth.

In order for glass-ceramics to be used for dental applications, they must possess high chemical durability, mechanical strength and toughness and should exhibit properties which mimic the natural tooth microstructure in order for it to be successful as an aesthetic. Glass-ceramics allow all these properties to be united within one material. For a glass-ceramic to have the desired properties, the glass is converted into a glass-ceramic via controlled crystallisation to achieve the crystal phase wanted and hence the desired properties it could possibly have. Hence, the glass-ceramic developed allows it to have properties such as low porosity, increased strength, durability, toughness etc. which are crucial in the field of dental restorations as it prevents restorative failures which are mainly due to stress and porosity which causes cracks and hence failures.

It took many years of research in order to get a material strong enough to be initially used as a dental reconstructive material. However over the past 10-15 years, research has progressed vastly and now glass-ceramics demonstrate good strength, high durability and good aesthetics. The development and processing of glass-ceramics has been focused on particular clinical applications, such as dental inlays, crowns, veneers, bridges and dental posts with abutments. Glass-ceramics are divided into seven types of materials:

- Mica glass-ceramics,

- Mica apatite glass-ceramics,

- Leucite glass-ceramics,

- Leucite apatite glass-ceramics,

- Lithium Disilicate glass-ceramics,

- Apatite containing glass-ceramics,

- ZrO_2-containing glass-ceramics.

The first commercially usable glass ceramic products for restorative dentistry were composites of mica glass ceramics. Dicor and Dicor MGC were products based on these. According to the mechanism of controlled volume crystallisation of glasses, tetrasilicic micas, $Mg_{2.5}Si_4O_{10}F_2$, showing crystal sizes of 1 to 2 μm in the glass ceramic were produced. Dicor being amongst them was shaped by means of centrifugal casting methods to produce dental restorations such as dental crowns and inlays. Depending on the different crystal sizes and the corresponding microstructure of the glass ceramic, it was also possible to manufacture glass ceramics for machining applications. Dicor MGC being amongst them. This resulted in the characteristic of good machinability in this type of glass-ceramic to be exploited and results concluded that crystals upto only 2 μm in length in the material improved mechanical strength over other materials.

Mica-apatite glass-ceramics have been produced in the SiO_2-Al_2O_3-Na_2O-K_2O-MgO-CaO-P_2O_5-F system. The main crystal phases are phlogopite, $(K, Na)Mg_3(AlSi_3O_{10})F_2$ and fluorapatite, $Ca_5(PO_4)_3F$. The base glass consists of three glass phases: a large droplet-shaped phosphate-rich phase, a small droplet-shaped silicate and a silicate glass matrix. Mica is formed during heat treatment, as in apatite-free glass-ceramics, by in-situ crystallization via the mechanism of volume crystallization. Apatite is formed within the phosphate-rich droplet phase. Astonishingly, every single apatite crystal possesses its own nucleation site in the form of a single phosphate drop. The glass-ceramic is biocompatible and suitable for applications in head and neck surgery as well as in the field of orthopaedics.

Leucite glass-ceramics can be formed by applying the advantage of the viscous flow mechanism. IPS Empress is of this type of glass-ceramic. The material is processed by using the lost wax technique, whereby a wax pattern of the dental restoration such as an inlay, onlay, veneer or crown is produced and then put in a refractory die material. Then the wax is burnt out to create space to be filled by the glass-ceramic. As the glass-ceramic has a certain volume of glass phase, the principle of viscous flow can be applied and hence the material can be pressed into a mould. Surface crystallisation and surface nucleation mechanisms were controlled in order for this type of glass-ceramic to be formed. Consequently, the manufacturing of inlays and crowns developed due to the application of viscous flow mechanism of glass-ceramics in different shapes. The resulting leucite glass-ceramic restorations transluceny, colour and wear resistance behaviour can then be adjusted to those of natural tooth. Additionally, the leucite glass-ceramic restorations can be produced by machining with CAD/CAM. IPS ProCAD and IPS Empress CAD are glass ceramics produced via this method. All leucite glass-ceramic restorations are bonded to the tooth structure with a luting material, preferably an adhesive bonding system. The retentive pattern produced on the glass-ceramic surface is particularly advantageous in this respect.

It was possible to develop a leucite apatite glass-ceramic derived from the SiO_2-Al_2O_3-Na_2O-K_2O-CaO-P_2O_5-F system by combining two different mechanisms, i.e. controlled surface nucleation and controlled bulk nucleation. IPS d.SIGN is amongst these. The glass-ceramic was prepared according to the classic method of glass-ceramic formation: melting, casting to prepare a glass frit, controlled nucleation and crystallization. A two-fold reaction mechanism leads to the precipitation of fluoroapatite, $Ca_5(PO_4)_3F$ and leucite, $KAlSi_2O_6$. SEM pictures show the two-phase crystal content of apatite and leucite in this type of glass-ceramic. Fluoroapatite phase takes the form of needle-shaped crystals whereas the oval areas are the leucite crystals. The clinical application of this glass-ceramic has been proven to be suitable for clinical application as veneering material on metal frameworks for single units as well as for large dental bridges involving more than three units.

The first glass-ceramic to be developed was by Stookey et al which contained Lithium disilicate. Further research into this field allowed for IPS Empress2 to be developed. This glass-ceramic was developed in order to extend the range of indications of glass-ceramics from inlay and crowns to three-unit bridges, by offering high strength, high fracture toughness and at the same time, a high degree of translucency. Both the flexural strength and fracture toughness of lithium disilicate glass-ceramics are almost three times of those of leucite glass-ceramics. Lithium disilicate glass-ceramic ingot is utilizied to produce the crown or bridge framework in combination with the viscous flow process. To further improve the aesthetic properties, i.e. translucency and shade match, and to optimally adjust the wear behaviour to that of the natural tooth, the lithium disilicate glass ceramic is veneered with an apatite-containing glass-ceramic using a sintering process.

In order to meet the demanding requirements of CAD/CAM applications, a lithium metasilicate glassceramic, IPS e.max was developed. This material, which is supplied in a typically blue colour, is adjusted by thermal treatment in order to demonstrate a characteristic tooth colour. The range of IPS e.max products also encompasses various apatite-containing glass ceramics that are suitable for both layering material on lithium disilicate glass-ceramic and veneering material on ZrO_2 sintered ceramic. The apatite crystal phase of the $Ca_5(PO_4)_3F$ type acts as a component that adjusts the optical properties of the restoration to natural tooth. For this reason, the crystallites are of nanoscale dimension.

ZrO_2 containing glass-ceramics was the first glass-ceramic developed to be fused to high strength ZrO_2 ceramic dental posts. The glass-ceramic contains $Li_{12}ZrSi_6O_{15}$ crystals as the main phase; however different types of crystals are also precipitated in the glassy matrix. ZrO_2 has become very interesting not only in the field of medicine but also in dental applications. High-strength and high toughness dental posts, crowns and bridges can be prepared from this material. In order for a dental restorative material to be of clinical success, their most important properties include high strength, high toughness, abrasion behaviour comparable to natural teeth, translucency, colour, durability) and the processing technologies (moulding, machining, sintering). Furthermore, the material should have good marginal fit with the tooth, biocompatibility, good mechanical properties and low porosity. In addition to the aforementioned properties, the recent requirement for dental restorative materials is for its appearance to be similar to that of a natural tooth.

Glass-ceramics have been researched immensely in order to fulfil high standards of function and aesthetics from an early stage. The trend for metal free dental restorations began from the 1970's whereby metal free feldspathic ceramics were reinforced with additional components. Since then, increasing the strength of these materials progressed rapidly by controlling the nucleation and crystallisation of glasses. These developments have now led to the introduction of a trend which is focused on achieving exceptional aesthetic results with glass ceramics as metal free dental restorations. Although glass-ceramics exhibit the desired properties for dental restoration, their main drawback is that they are brittle which the main cause of failure is. This is due to either fabrication defects; which are created during production of the glass-ceramic or secondly, surface cracks; which are due to machining or grinding. Therefore when processing the glass-ceramic, care needs to be taken in addition to choosing the suitable method for production for specific compositions of the glass-ceramic in order to improve their mechanical properties.

Apart from the use of glass-ceramics for dental restorations, they can also be applied as bioactive materials for the replacement of hard tissue. Bone is a complex living tissue which has an elegant structure at a range of different hierarchical scales. It is basically a composite comprising collagen, calcium phosphate (being in the form of crystallised hydroxyapatite, HA or amorphous calcium phosphate, ACP) and water. Additionally, other organic materials, such as proteins, polysaccharides, and lipids are also present in small quantities. Because bone is susceptible to fracture; there has always been a need, since the earliest time, for the repair of damaged hard tissue. Many years of research has attempted to use biomaterials to replace hard tissue, ranging from using bioinert materials, to bioactive materials such as 'Bioglass' to 'Apatite-wollastonite (A-W) glass-ceramics and to calcium phosphate materials. Calcium phosphate based materials have received a great deal of attention in this field due to their similarity with the mineral phase of bone.

Kinetic Theory of Glass Formation

The crystallization of a melt consists of two processes (i) nucleation of the crystalline phase from the melt (ii) growth of these nuclei. As the free energy of the crystal is lower than that of the melt below the melting point, the crystallization is thermodynamically favourable. However, there are two barriers to this transformation. The first barrier arises from the increase in the surface energy that accompanies the creation of a new solid-liquid interface as a nucleus forms. This results in a thermodynamic barrier to nucleation. The second barrier is the kinetic barrier that comes from the difficulty in diffusion of the material through the liquid to the nucleus during its formation and growth. This becomes more and more difficult as the viscosity of the melt increases during cooling. If these barriers can be overcome, crystallization would result; otherwise a glass would form. If a relation can be derived for the rate of crystal formation in terms of the various parameters, then one can predict what factors would lead to a low rate and so lead to glass formation. For this we need to derive a relation for the rate of nucleation and also for the rate of crystal growth.

Nucleation

First let us consider the process of nucleation. In it a number of structural units in the liquid come together by diffusing through the liquid to form a crystalline cluster. This cluster is called an embryo. These structural units can be single atoms as in the case of a metal or they can be formula units such as SiO_4 in case of silica. As the basic formula units in the liquid come together to form a crystalline embryo, an interface is created between the embryo and the surrounding liquid. This interface has a certain surface energy. In the beginning when the embryo is small, the increase in the surface energy due to the formation of the embryo is larger than the decrease in the volume free energy due to the liquid → crystal transformation because the surface to volume ratio of an embryo is large due to its very small size. The embryo formation therefore causes an increase in energy and the embryos tend to dissolve back. At any time there would be a distribution of sizes of the embryos in the liquid. As the embryo size increases, the decrease in the volume free energy approaches the increase in the surface energy until at some size, the two are equal in magnitude. This size of the embryo is the critical size because further addition of a single formula unit results in a net decrease in the free energy and the embryo is now stable. Embryo has now become a nucleus and this size is called the critical nucleus size. This critical nucleus size can be derived as follows.

If the thermodynamic barrier for the formation of a nucleus is ΔG^* and the kinetic barrier is ΔG_D, then the rate of nucleation can be expressed as:

$$I = Ae^{\left[-\left(\Delta G^* + \Delta G_D\right)/kT\right]}$$

where A is a pre exponential factor which can be approximated as:

$$A = n_v \left(kT/h\right)$$

Where n_v is the number of formula units of the crystallising component phase per unit volume of the melt and h is the Planck's constant.

An expression for ΔG^* is now to be derived. Let the change in the free energy per unit volume between the liquid and the crystal be ΔG_v. Note that ΔG_v is a negative quantity because crystal is the more stable phase. Let σ be the surface energy per unit area of the liquid-solid interface. For the formation of a spherical embryo of radius r,

- Change in the volume free energy = $\Delta G_v (4/3)\pi r^3$.

- Change in the surface energy = $4\pi r^2 . \sigma$.

- Total change in energy $\Delta G = \Delta G_v (4/3)\pi r^3 + 4\pi r^2 . \sigma$.

The variation of the two terms on the right hand side and of the total energy with radius is shown in figure. As r increases, the surface energy term dominates at small r and the total free energy increases at first; after a critical value of r, the volume free energy term prevails and the total free energy decreases as r is further increased. The critical value of r, r* is at the maximum in the free energy curve and is obtained by equating the derivative of ΔG to zero:

$$\frac{\partial \Delta G}{\partial r} = 4\pi r^2 \Delta G_v + 8\pi r\sigma$$

At r = r*, $\dfrac{\partial \Delta G}{\partial r} = 0$ or $r = r^* = -\dfrac{2\sigma}{\Delta G_v}$

Embryos smaller than r* are unstable. The free energy barrier associated with the formation of a critical sized nucleus is obtained by putting r = r*.

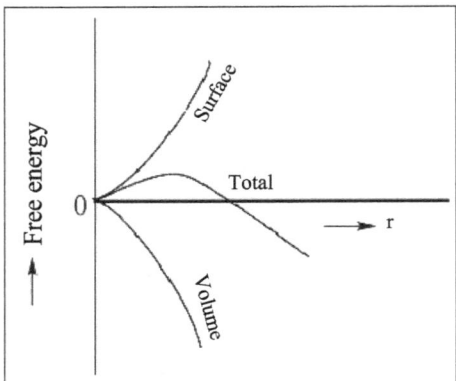

Changes in the volume free energy, the surface energy and the total free
energy terms during the nucleus formation.

$$\Delta G^* = \frac{16\pi\sigma^3}{3\Delta G_v^{\,2}}$$

The kinetic barrier to diffusion, ΔG_D can be estimated in the following manner. The diffusion coefficient D in the liquid can be written in terms of this barrier as:

$$D \sim \exp-\left(\Delta G_D / kT\right)$$

The pre exponential is given by $(kT\delta^2 / h)$ where δ is the atomic jump distance. Putting this value:

$$D = \left(kT\delta^2 / h\right) \exp-\left(\Delta G_D / kT\right)$$

In a liquid the diffusion coefficient can also be expressed using the Stokes-Einstein relation:

$$D = kT / 3\pi\eta\delta$$

Combining $D \sim \exp-\left(\Delta G_D / kT\right)$ and $D = kT / 3\pi\eta\delta$ we get,

$$\exp-\left(\Delta G_D / kT\right) = \left(h / 3\pi\eta\delta^3\right)$$

Using equation $\exp-\left(\Delta G_D / kT\right) = \left(h / 3\pi\eta\delta^3\right)$ in eq. $I = Ae^{\left[-\left(\Delta G^* + \Delta G_D\right)/kT\right]}$ we get an expression for the rate of nucleation as:

$$I = (n_v kT / 3\pi\eta\delta^3) \exp(-\Delta G^* / kT)$$

Or $I = \left(n_v kT / 3\pi\eta\delta^3\right) \exp\left(-16\pi\sigma^3 / 3\Delta G_v^{\,2} kT\right)$

Note that in this expression, the thermodynamic effect is provided by the exponential term while the kinetic effect is provided by the pre exponential term through the viscosity, η.

In the above discussion we have considered nucleation from within the melt. This is called homogeneous nucleation. The nuclei can also grow on surfaces which are in contact with the molten liquid. The walls of the container obviously provide one such surface. In addition there may be foreign particles of a higher melting point material within the melt. The surfaces of these particles can also act as nucleation sites. This is called heterogeneous nucleation. The heterogeneous nucleation is in fact easier than the homogeneous nucleation if the surface is wetted by the liquid because then the effective surface energy is lower. Figure shows a crystal embryo in the form of a cap forming on a surface. The contact angle is θ. In this case, the free energy barrier for the formation of a critical sized nucleus cap is given by:

$$\Delta G_{het}^{*} = \Delta G_{homo}^{*} . f(\theta)$$

where $f(\theta) = (2 - 3\cos\theta + \cos^3\theta)/4$.

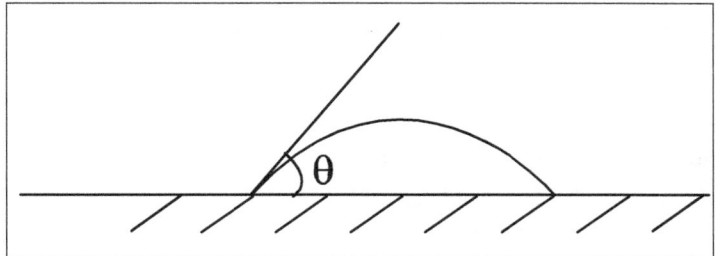

Formation of a nucleus in the form of a cap on a surface in the case of the heterogeneous nucleation.

If there is good wetting between the crystalline phase and the surface, then θ is low and $f(\theta) < 1$. Thus the barrier for heterogeneous nucleation is lower than that for homogeneous nucleation. The expression for the nucleation rate in this case is obtained by using this value of ΔG^{*} in Equation.

Let us now discuss the effect of supercooling below the melting point in light of the eq. $I = (n_v kT / 3\pi\eta\delta^3) \exp(-16\pi\sigma^3 / 3\Delta G_v^2 \, kT)$ and eq. $r = r^{*} = -\dfrac{2\sigma}{\Delta G_v}$. Just below the melting point, ΔG_v is very small and r^* is very large. From eq. $I = (n_v kT / 3\pi\eta\delta^3) \exp(-16\pi\sigma^3 / 3\Delta G_v^2 \, kT)$, or, equivalently due to the large critical size required, the nucleation rate is nearly zero. As the temperature decreases, the critical size required for a stable nucleus decreases and their number increases; the nucleation rate increases. If the viscosity does not fall too rapidly, the nucleation rate continues to increase. At still lower temperatures, the viscosity increases and the nucleation rate begins to decrease. This is shown in figure.

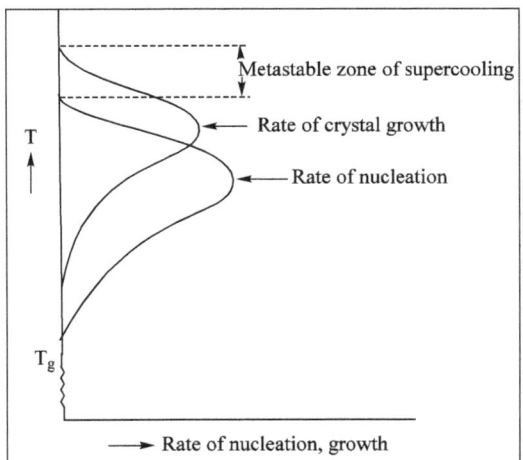

Changes in the nucleation rate and the growth rate on cooling below the melting point.

Growth

Once a stable nucleus has formed it grows by addition of atoms from the liquid. Expression for the growth rate is derived using the following picture of the growth process. It is assumed that the growth is occurring uniformly over all the surface of the nucleus. An atom from the liquid has to jump across the liquid-solid barrier and get incorporated in the growing crystal. There is also the reverse process of atoms from the solid making the reverse jump but the probability of this is smaller due to an increase in the free energy in the reverse process. Using this picture, the following expression for the growth rate is derived.

$$U = \frac{kT}{3\pi a_o^2 \eta}\left[1 - \exp\left(\Delta G / kT\right)\right]$$

Here ΔG is the thermodynamic barrier for crystal growth and η as before comes from the kinetic barrier, a_o is interatomic separation distance, v is the vibrational frequency. As in the case of the nucleation rate, the exponential factor decreases as the temperature is decreased below the melting point because the thermodynamic barrier decreases. The quantity in the brackets therefore increases so the growth rate increases with increasing supercooling below the melting point. At some temperature, the viscosity increases so much that the pre exponential factor begins to dominate and the growth rate begins to decrease after reaching a maximum, just like the nucleation rate. In case of nucleation, the nucleation rate does not become appreciable until a certain degree of supercooling is reached. So there is a temperature range below the melting point in which the nucleation rate is zero. In contrast, the growth rate can be finite just below the melting point because there may be present some nuclei due to homogeneous or heterogeneous nucleation. This is also shown in figure.

Volume Transformed at Any Time

At any temperature below the melting point, nucleation and growth occur simultaneously. At any temperature, a certain volume fraction would have transformed to the crystalline phase. If V is the total volume and V_x is the volume of the crystalline phase at any instant, then this volume has been shown to be approximately given by:

$$\frac{V_x}{V} = 1 - \exp\left(-\frac{\pi}{3}IU^3t^4\right)$$

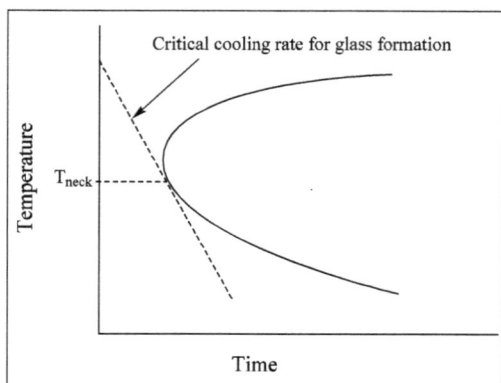

The time-transformation-temperature (TTT) curve for a glass forming melt for an arbitrary volume fraction of glass, say 10^{-6}. The cure is C shaped with a neck at T_{neck}.

Here t is the time the sample is held at a temperature below the melting point. The time required to crystallize a given constant fraction, say 10 %, at any temperature can be calculated from equation $\frac{V_x}{V} = 1 - \exp\left(-\frac{\pi}{3}IU^3t^4\right)$. This time is very large just below the melting point, decreases as the temperature is decreased and then again increases as the temperature is further lowered. We can defined the solidified material to be a glass if it contains less than a certain amount of the crystalline phase, say $(V_x/V) < 10^{-6}$. If the above calculation is done for this volume fraction (or any other volume fraction) a curve as shown in figure is obtained. This is called the time-temperature-transformation (TTT) curve.

Kinetic Criterion for Glass Formation

As can be seen, the TTT curve is C shaped with a neck at a certain temperature, T_{neck}. We can draw a straight line from the melting temperature such that it is tangent to this nose. The slope of this line gives a cooling rate. If the melt is cooled at this or a lower cooling rate, then no crystal formation would occur and a glass would result. For the oxides which form glasses this critical cooling rate is lower than the rate corresponding to the normal cooling so that such oxides form glass easily on normal cooling. On the other hand, the critical cooling rate may be as high as 10^{12} °C. sec^{-1} for pure metals, a very difficult to achieve rate. What factors would lead to a low critical cooling rate and so to glass formation? Some of these are listed below in light of the ideas discussed above:

- Minimization of heterogeneous nucleation by elimination of impurities such as refractory particles and minimization of the contact with the container walls. Superheating of the melt would ensure complete melting of the last solid constituent which might act as heterogeneous nucleus. Use of fluxes such as PbO which forms low melting compounds with many oxides and thus eliminate them as potential heterogeneous nucleation sites would also favour glass formation.

- The viscosity should be high at the melting point or should increase rapidly on cooling below the melting point. A high viscosity presents a high kinetic barrier to crystallization.

- If the melt contains many different elements which have to rearrange themselves to form the crystalline structure, the crystal growth would be difficult and so the critical cooling rate would be low. This is the reason that the glass forming compositions are generally very complex.

Structural Theories of Glass Formation

Before starting with this theory, it is necessary to understand the Goldschmidt's radius ratio criterion for the formation of oxide crystal structures.

Goldschmidt's Radius Ratio Criterion

Before considering the Goldschmidt' radius ratio criterion it would be appropriate to consider briefly Pauling's rules on grouping of atoms in crystals with largely ionic bonding. In the ionic crystals having predominantly ionic bonding, the energy can be minimized if the cations and the anions are packed relative to each other in such a way that the electrostatic attractive energy is maximized and the repulsive energy is minimized. Pauling has made certain generalizations which predict the majority of the known ionic crystal structures. These generalizations have come to be known as Pauling's rules. According to the first of these rules, the anions surround each cation to form an anion polyhedron. The cation-ion distance is the sum of their radii. Here one should understand that the concept of the radius of an ion is strictly empirical. The justification for this concept is that using these values, the interionic distances between the crystals can be predicted within a few percent using a self-consistent set of ionic radii. A different set of ionic radii is to be used for different coordination numbers. The anions are thus assumed to be touching the cations. The number of anions surrounding the cation is called the coordination number. This number depends on the ratio of the radius of the cation to the radius of the anion. For example, consider a coordination number of 3 i.e. three anions forming a triangle with the cation sitting at the centre of the triangle. For the cation to touch the anions, the radius ratio must be larger than a critical value as shown in figure.

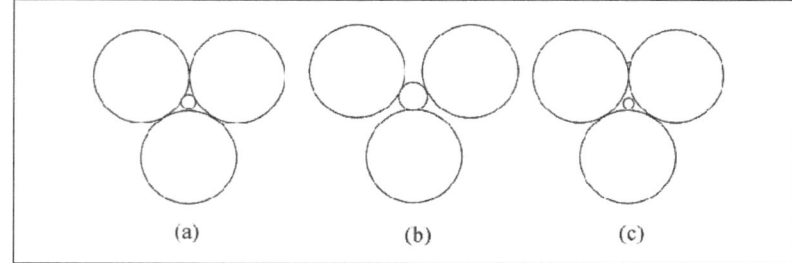

Anions in a triangular arrangement around the cation (a) r_c/r_a is just sufficient for the cation to touch the anions (b) if r_c/r_a is larger than the critical value in (a), then the cation continues to touch all the anions (c) if r_c/r_a is smaller than the critical value, the cation cannot touch the anions.

The values of the critical radius ratios for the different coordination are given in table. The radius ratio places an upper limit on the coordination of the cation. Thus for a radius ratio of 0.4, the coordination number cannot be more than 4. It is possible to have a smaller coordination than this upper limit. But the energy of the resulting configuration will be higher – it can be reduced by increasing the coordination to maximum permitted because the electrostatic energy of an array is decreased as a larger number of oppositely charged ions are brought in contact.

Table: Critical radius ratios for different coordination's.

Coordination number	Disposition of the anions around the central cation	Critical radius ratio
8	Cube corners	≥ 0.732
6	Octahedron corners	≥ 0.414
4	Tetrahedron corners	≥ 0.225
3	Triangle corners	≥ 0.155
2	Linear	≥ 0

It should also be noted that sometimes it is possible to have a higher coordination than permitted by the radius ratio. In this case, for the anions to touch the central cation, some deformation of the ions will be necessary. This would increase the energy of the configuration. However, if the cation is highly charged, than the reduction in the electrostatic energy due to increased number of anions surrounding the cation may be sufficient to compensate for this increase in energy.

The second rule of Pauling concerns the electrical neutrality. The structure must be neutral not only on the macroscopic scale but also at the atomic level. In a structure a cation may be surrounded by anions having different charges. Thus in Al_2O_3, the Al^{3+} is surrounded by 6 O^{2-} while in kaolinite it is surrounded by $4OH^-$ and 3 O^{2-}. It is therefore not straightforward to determine if the electrical neutrality condition is satisfied. To determine if the requirement of electrical neutrality is satisfied in a case, the concept of bond strength is introduced. The strength of an ionic bond donated from an anion to a cation is defined as the formal charge on the cation divided by its coordination number. Thus the charge on Si is +4 and it has a tetrahedral coordination. So its bond strength is 4/4 =1. It should be noted that in the above definition only the number of anions surrounding the cations is counted – it does not matter if they are different. The second rule of Pauling states that in a stable structure the total strength of the bonds reaching from all the anions surrounding the cations must be equal to the charge on the cation.

Pauling's third rule is about how the polyhedra are connected to each other. According to this rule, the polyhedra must be linked at the corners rather than the edges or at the face. This is because the separation between the cations increases as the cations are linked at the corners, at the edges and at the faces. To minimize the electrostatic repulsion therefore, the polyhedra must be linked at the corners. This is especially true for the polyhedra formed of cations with a high charge and low coordination number.

Now we come back to the Goldschmidt's radius ratio criterion for glass formation. Goldschmidt observed that oxides of the type A_nO_m form a glass most easily when the radius ratio of the cation to anion (oxygen), r_A/r_O, is between 0.2 to 0.4. In inorganic compounds, this ratio determines the structure of the oxide. A radius ratio between 0.2 to 0.4 implies that the structure consists of tetrahedral coordination with four anions surrounding each oxygen. On this basis, Goldschmidt came to the conclusion that only those melts which contained tetrahedrally coordinated oxygen formed glass on cooling. This conclusion was purely empirical. Although largely true, there are many exceptions. Thus 9TeO.PbO form glass but the coordination number of Te is 6. On the other hand, BeO with r_{Be}/r_O = 0.221 does not form a glass. It should also be noted that the bonding is not purely ionic in the glass forming oxides while the radius ratio rule is derived for ionic bonds.

Zachariasen's Criterion for Glass Formation

The next idea on glass formation came from Zachariasen. Zachariasen's ideas were addressed to the question of glass formation but they became the foundation on which the models of the glass structure were subsequently built and form an important milestone in the glass science. Zachariasen's idea grew out of his observation that the silicate crystals which readily form glasses on melting and cooling have a three dimensional network structure. Since the density and the mechanical properties of the glass and the crystalline forms are similar, it occurred to Zachariasen that the glass must also have a three dimensional network structure but the networks are not symmetrical and periodic as in crystals. The ability to form a network was therefore the criteria he proposed to predict whether an oxide would form a glass or not. Zachariasen then considered the structural arrangements that would be necessary to form a network structure. Based on this he proposed that the glass formation may occur in an oxide if the following conditions are met.

- Each oxygen ion should be linked to not more than two cations (otherwise, the linkage distortion to give a random network structure would not be possible; also there is a need to keep the cations as far apart as possible).

- The coordination number of the oxygen ions about the glass forming cation must be small, four or less (otherwise an octahedral or cubic unit will result, with no flexibility).

- The oxygen polyhedra must share corners with each other, not the edges or the faces (to keep the cations as far apart as possible).

- At least three corners of a polyhedron must be shared (necessary to form a network).

As an example consider the case of silica, SiO_2. For this case the radius ratio is,

$$\frac{r\left(Si^{4+}\right)}{r\left(O^{2-}\right)} = \frac{0.40}{1.40} = 0.29$$

According to table, SiO_2 prefers tetrahedral bonding and so satisfies Zachariasen's second condition. Next we find the coordination number for oxygen. This is done by using the condition that the ratio of the charge to coordination number should be same for both the ions, i.e.

$$\frac{charge(Si^{4+})}{CN(Si^{4+})} = \frac{4}{4} = \frac{charge(O^{2-})}{CN(O^{2-})} = \frac{2}{CN(O^{2-})}$$

So, that $CN(O^{2-}) = 2$. This satisfies the first condition of Zachariasen. Thus SiO_4 tetrahedra, linked at corners will form a glass. Based on these rules, Zachariasen predicted that the following oxides should form a glass:

$$B_2O_3, SiO_2, GeO_2, P_2O_5, As_2O_3, Sb_2O_3, V_2O_5, Sb_2O_5, P_2O_3, As_2O_5, Nb_2O_5, \text{ and } Ta_2O_5.$$

Out of these only the first seven can be made into a glass under ordinary conditions. Moreover, in addition subsequently the following oxides have also been prepared in the glass form:

$$In_2O_3, Tl_2O_3, SnO_2, PbO_2, SeO_2.$$

It should be noted that Zachariasen only proposed that an oxide may form a glass if certain conditions are met. He proposed the random network based on the similarity of the structure and the internal energy between the crystalline and glassy form of the oxides. Although proposed as a test for the glass forming ability, these ideas have become the basis for glass structure.

Some other Criteria for Glass Formation

Smekal proposed that for glass formation, the bonding in the material should be neither fully ionic nor fully covalent but should be of mixed nature. The covalent bonds are highly directional and have rigid bond angles. A purely covalent bonding therefore cannot produce a random network structure characteristic of the glass. On the other hand, the ionic bonds are not directional and are incapable of forming a network. The bonding in a glass must therefore be of a mixed character. According to Smekal this requirement can be met in the following materials:

- Inorganic compounds in which the bonding is partly covalent and partly ionic. Examples are SiO_2, B_2O_3.

- Elements such as S, Se which have a chain structure. The bonding within the chain is by covalent bonds, but the chains are held together by van der Waals bonds.

- Polymers with large molecular chains. Here again the bonding within the chain is by covalent bonds, but the chains are held together by van der Waals bonds.

Taking this idea further, Stanworth grouped the oxides into three groups on the basis of the electronegativity of the cation. Since the anion is oxygen in all cases, the electronegativity of the cation is a measure of the ionic character of the bond, higher electronegativity implying a lower ionic character. Table below groups the elements on the basis of their electronegativity.

Table: Grouping of elements in oxide glasses on the basis of electronegativity.

Group I		Group II		Group III	
Boron	2.0	Beryllium	1.5	Magnesium	1.2
Silicon	1.8	Aluminium	1.5	Calcium	1.0
Phosphorus	2.1	Titanium	1.6	Strontium	1.0
Arsenic	2.0	Zirconium	1.6	Barium	0.9
Antimony	1.8	Tin	1.7	Lithium	1.0
				Sodium	0.9
				Potassium	0.8

- Cations in Group I form bonds with oxygen with ionic character greater than 50%. They should be network formers and should form glass by themselves. They are known as network formers.

- Cations in Group II have a slightly lower electronegativity. They cannot form glass by themselves but can replace some cations in Group I. They are called intermediates.

- Lastly, the cations in Group III with still lower electronegativity form highly ionic bonds with oxygen. They cannot form network. They can act only as network modifiers if added to the network forming oxides. These oxides are called modifiers.

Dietzel considered the direct Coulombic attraction between the cation and the anion. This attractive force is given by,

$$\text{Attractive force} = \frac{(Z_c e).(Z_a e)}{(r_c + r_a)^2}$$

Here, Z_c, Z_a are the number of charges on the cation and the anion respectively, r_c and r_a are the radii of the cation and the anion respectively and e is the electronic charge.

For oxides, Z_a is fixed at 2. If $r_c + r_a = R$, then Dietzel defined the field strength as:

Field strength, $F_s = Z_c/R^2$.

A high field strength implies a high attractive energy between the cation and the anion. Based on empirical observations, Diezel related the field strength to the ability for glass formation. He proposed the following:

- For glass formers the field strength is > 1.3.

- For intermediates, the field strength is 0.4 and 1.3.

- For modifiers the field strength is less than 0.4.

Sun considered in a similar manner, the single bond strength of the oxides as a criterion for glass formation. He defined the single bond strength as the dissociation energy of the oxide divided by the coordination number. He then proposed that:

- If the single bond strength (SBS) is high, > 80 Kcal/bond then the element is a glass former.

- If the SBS is between 60 to 80 Kcal/bond, the element is an intermediate.

- If SBS < 60 Kcal/bond, the element is a modifier.

Applications of Glass Ceramics

Construction Application

In the last few decades, there has been considerable research on the production of glass-ceramics from silicate waste, such as coal combustion ash, slag from steel production, Fly Ash and filter dusts from waste incinerators, mud from metal hydrometallurgy, etc. These low-cost glass-ceramics recycled from industry silicate waste are generally strong, hard and chemically resistant. Their intended use is for abrasion and chemically resistant parts or floor and wall tile. One of the most popular glass–ceramic used in construction is Neopariés LT, with wollastonite as the main crystalline phase. Neopariés glass–ceramic panels are an ideal alternative to stone for interior and exterior applications.

Optical Application

In optical field, many Glass–Ceramics show high translucency or even can be transparent because of the fact that zero porosity can be relatively easily achieved. These make Glass–Ceramics excellent material for optical applications. For instance, transparent and low thermal expansion Glass–Ceramics based on lithium aluminosilicate (LAS) system have been used as telescope mirror blanks and laser gyroscopes. Glass-ceramics can demonstrate optical properties similar to those of single crystals, meanwhile, glass-ceramics tend to be more cost efficient and more suitable for manufacturing larger objects with intricate shapes. Moreover, glass-ceramics doped with transition metal ions were developed for use in broadband optical amplification, tunable and infrared lasers, phosphor with tunable UV/blue luminescence behavior, and in solar collectors.

Military Application

In military field, Glass–Ceramics now are used in nosecones of high–performance aircraft and missiles. Materials used in these applications must exhibit a challenging combination of properties to withstand critical conditions resulting from high–speed flying in the atmosphere: Low coefficient of thermal expansion; high mechanical strength; high abrasion resistance; high radar wave transparency for navigation. However, little has been published and patented on the chemical composition, microstructure, and preparation processes of the glass-ceramics used for military, because of the sensitive nature of this military-related research. No glass, metal or single crystal can simultaneously meet all of these relevant specifications, while glass–ceramics with the tailored properties were able to achieve the challenge. Another example is the view-window for armored vehicles or tank, which requires that material of the view-window to be both strong and highly-translucent. Glass-ceramics with high strength, high toughness, and high translucency are excellent candidates for these particular uses.

Medical Application

In medical field, bioglass, for instance the "gold standard" bioglass 45S5 invented by Larry Hench, has been successfully used in medical field. However, the inherent low strength and low toughness limit the application of bioglass as load-bearing biomaterial. With crystalline phases as strengthening and toughening phases, Glass–Ceramics overcome the weakness of bioglass. Moreover, glass-ceramics generally show better bioactivity and biocompability than sintered ceramics due to the presence of glass phase in glass-ceramics. For instance, A–W glass–ceramic that contains apatite and β–wollastonite ($CaO \cdot SiO_2$) crystals (with the commercial brand name of Cerabone) is considered as the most outstanding bioactive glass–ceramics for hard tissue repair. Research on improving the properties of glass-ceramic biomaterials are still going on and one direction is to introduce greater multifunctionality to glass-ceramics through inorganic modifications, for instance, bioactivity and antimicrobial activity can be simultaneously enhanced by incorporation of specific ions, such as Cu^{2+}, Sr^{2+}, and Zn^{2+} ions.

Electronic Application

In electronic field, all–solid–state secondary batteries with inorganic solid electrolytes are expected to be next–generation high–output batteries. The crucial requirements for appropriate solid electrolytes are high ionic conductivity and good formability. Different types of inorganic solid electrolytes made by glass–ceramics have been developed, for instance, glass–ceramics has the crystalline form of $Li_{1+x+y}Al_xTi_{2-x}Si_yP_{3-y}O_{12}$ exhibited a high lithium–ion conductivity of 10^{-3} S·cm^{-1}. Compared with liquid electrolytes, solid electrolytes made by glass-ceramics is more stable in the open atmosphere and even to exposure to moist air, thus, they are expected to be applied for various uses.

Kitchenware Application

In kitchenware field, higher toughness (compared with glass), appealing aesthetics, and very low thermal expansion coefficient make glass–ceramics the excellent material for kitchenware, such as cooktops, cookware, and bakeware. In the late 1960s, Corning introduced the concept of cooking on a smooth white glass-ceramic. To date, many glass-ceramics with different compositions have been developed to use as kitchenwares. The most widely used system is the $Li_2O-Al_2O_3-SiO_2$ (LAS) system with additional components, such as CaO, MgO, ZnO. The main crystalline phase is a β–quartz solid solution, which has an overall negative CTE. LAS Glass–Ceramics can sustain repeated and quick temperature changes of 800 to 1000 °C.

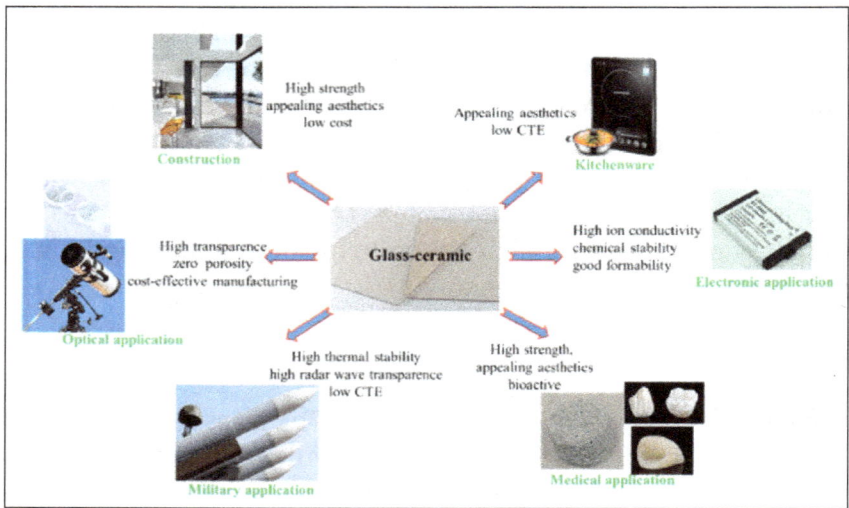

Applications of glass–ceramics in a wide range of fields.

References

- Glass-ceramics: ukdiss.com, Retrieved 23, April 2020

- Kinetic-theory-of-glass-formation, Ceramics: inflibnet.ac.in, Retrieved 14, June 2020

Magnetic Ceramics

Magnetic ceramics are the oxide materials made up of ferrites, which are composed of iron oxide in combination with some other metal. They are crystalline minerals that exhibit ferrimagnetism. It is a type of permanent magnetization. This chapter discusses in detail the different concepts related to magnetic ceramics such as magnetic moment, magnetization, etc.

Magnetic ceramics are important materials for a variety of applications such data storage, tunnel junctions, spin valves, high frequency applications etc. These materials possess extra-ordinary properties such as strong magnetic coupling, low loss characteristics and high electrical resistivity which is often related to their structure and composition. Depending upon the type of application, based on the knowledge of materials, one can choose appropriate material.

To build an understanding, we will first have a brief look at the basics of magnetic properties, followed by various types of magnetism in materials and key characteristics and differences. This will be followed by a discussion on magnetic ceramics with emphasis on some of the key applications.

Magnetic Moment

The magnetic moment of a magnet is a quantity that determines the torque it will experience in an external magnetic field. A loop of electric current, a bar magnet, an electron, a molecule, and a planet all have magnetic moments.

The magnetic moment may be considered to be a vector having a magnitude and direction. The direction of the magnetic moment points from the south to north pole of the magnet. The magnetic field produced by the magnet is proportional to its magnetic moment. More precisely, the term *magnetic moment* normally refers to a system's magnetic dipole moment, which produces the first term in the multipole expansion of a general magnetic field. The dipole component of an object's magnetic field is symmetric about the direction of its magnetic dipole moment, and decreases as the inverse cube of the distance from the object.

Definition

The magnetic moment is defined as a vector relating the aligning torque on the object from an externally applied magnetic field to the field vector itself. The relationship is given by:

$$\tau = \mu \times B$$

where τ is the torque acting on the dipole and B is the external magnetic field, and μ is the magnetic moment.

This definition is based on how one would measure the magnetic moment, in principle, of an unknown sample.

Units

The unit for magnetic moment is not a base unit in the International System of Units (SI). As the torque is measured in newton-meters (N·m) and the magnetic field in teslas (T), the magnetic moment is measured in newton-meters per tesla. This has equivalents in other base units:

$$N \cdot m/T = A \cdot m^2 = J/T$$

where A is amperes and J is joules.

In the CGS system, there are several different sets of electromagnetism units, of which the main ones are ESU, Gaussian, and EMU. Among these, there are two alternative (non-equivalent) units of magnetic dipole moment:

$$1 \text{ statA} \cdot cm^2 = 3.33564095 \times 10^{-14} \text{ A} \cdot m^2 \text{ (ESU)}$$

$$1 \text{ erg/G} = 1 \text{ abA} \cdot cm^2 = 10^{-3} \text{ A} \cdot m^2 \text{ (Gaussian and EMU)},$$

where statA is statamperes, cm is centimeters, erg is ergs, G is gauss and abA is ab-amperes. The ratio of these two non-equivalent CGS units (EMU/ESU) is equal to the speed of light in free space, expressed in $cm \cdot s^{-1}$.

All formulae in this article are correct in SI units; they may need to be changed for use in other unit systems. For example, in SI units, a loop of current with current I and area A has magnetic moment IA, but in Gaussian units the magnetic moment is IA/c.

Two Representations of the Cause of the Magnetic Moment

The preferred classical explanation of a magnetic moment has changed over time. Before the 1930s, textbooks explained the moment using hypothetical magnetic point charges. Since then, most have defined it in terms of Ampèrian currents. In magnetic materials, the cause of the magnetic moment are the spin and orbital angular momentum states of the electrons, and varies depending on whether atoms in one region are aligned with atoms in another.

Magnetic Pole Representation

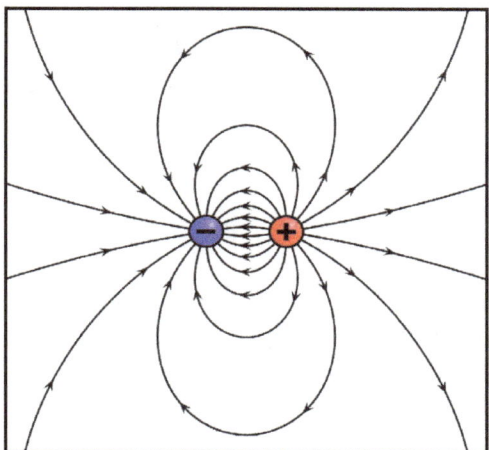

An electrostatic analog for a magnetic moment: two opposing charges separated by a finite distance.

The sources of magnetic moments in materials can be represented by poles in analogy to electrostatics. Consider a bar magnet which has magnetic poles of equal magnitude but opposite polarity. Each pole is the source of magnetic force which weakens with distance. Since magnetic poles always come in pairs, their forces partially cancel each other because while one pole pulls, the other repels. This cancellation is greatest when the poles are close to each other i.e. when the bar magnet is short. The magnetic force produced by a bar magnet, at a given point in space, therefore depends on two factors: the strength p of its poles (magnetic pole strength), and the vector l separating them. The moment is related to the fictitious poles as

$$\mu = p\mathbf{l}.$$

It points in the direction from South to North pole. The analogy with electric dipoles should not be taken too far because magnetic dipoles are associated with angular momentum. Nevertheless, magnetic poles are very useful for magnetostatic calculations, particularly in applications to ferromagnets. Practitioners using the magnetic pole approach generally represent the magnetic field by the irrotational field H, in analogy to the electric field E.

Integral Representation

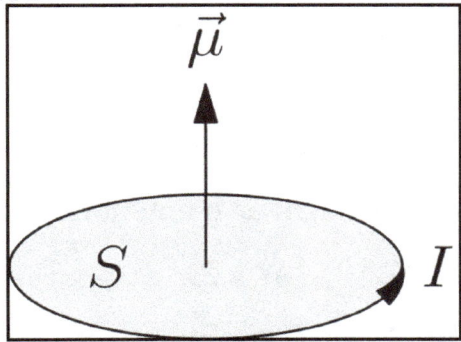

Moment $\boldsymbol{\mu}$ of a planar current having magnitude $\mathbf{I}$ and enclosing an area $\mathbf{S}$.

We start from the definition of the differential magnetic moment pseudovector:

$$\mu = \tfrac{1}{2}\mathbf{r} \times \mathbf{j}$$

where × is the vector cross product, r is the position vector, and j is the electric current density. It is very similar to the differential angular momentum, defined as:

$$\mathbf{l} = \mathbf{r} \times (\rho\mathbf{v})$$

where ρ is the mass density and v is the velocity vector. Like in every pseudovector, by convention the direction of the cross product is given by the right hand grip rule. Practitioners using the current loop model generally represent the magnetic field by the solenoidal field B, analogous to the electrostatic field D.

The integral magnetic moment of a charge distribution is therefore:

$$M = \tfrac{1}{2}\iiint_{v} \mathbf{r} \times \mathbf{j}\, dV,$$

Let us start with a point particle; in this simple situation the magnetic moment is:

$$M = \tfrac{1}{2}q\mathbf{r} \times \mathbf{v},$$

wherer is the position of the electric chargeq relative to the center of the circle and v is the instantaneous velocity of the charge, giving an electric current density j.

On the other hand, for a point particle the angular momentum is defined as:

$$\mathbf{L} = \mathbf{r} \times \mathbf{p} = m\mathbf{r} \times \mathbf{v},$$

and in the planar case:

$$M = \tfrac{1}{2}\iint_{s} \mathbf{r} \times \mathbf{j}\, dS,$$

by defining the electric current with a vector areaS (the x-, y-, and z-coordinates of this vector are the areas of projections of the loop onto the yz-, zx-, and xy-planes):

$$M = \tfrac{1}{2}I\int_{\partial S} \mathbf{r} \times d\mathbf{r}.$$

Then by Stokes' theorem, integral magnetic moment then becomes expressible as:

$$M = I\mathbf{S}.$$

The factor 1/2 in our definition above is only due to historical reason: the old definition of the magnetic moment was this last integral equation. If one had started from a differential definition:

$$\mu = \mathbf{r} \times \mathbf{j}.$$

then the coherent integral expression would have been:

$$M = 2I\mathbf{S}.$$

Magnetic Moment of a Solenoid

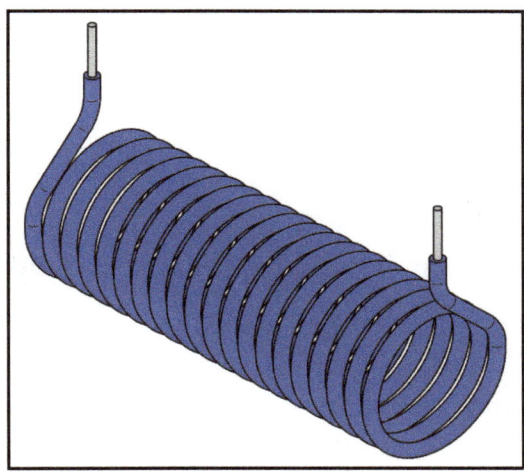

Image of a solenoid.

A generalization of the above current loop is a coil, or solenoid. Its moment is the vector sum of the moments of individual turns. If the solenoid has N identical turns (single-layer winding) and vector area S,

$$\mu = NI\mathbf{S}.$$

Magnetic Moment and Angular Momentum

The magnetic moment has a close connection with angular momentum called the gyromagnetic effect. This effect is expressed on a macroscopic scale in the Einstein-de Haas effect, or "rotation by magnetization," and its inverse, the Barnett effect, or "magnetization by rotation." In particular, when a magnetic moment is subject to a torque in a magnetic field that tends to align it with the applied magnetic field, the moment precesses (rotates about the axis of the applied field). This is a consequence of the concomitance of magnetic moment and angular momentum, that in case of charged massive particles corresponds to the concomitance of charge and mass in a particle.

Viewing a magnetic dipole as a rotating charged particle brings out the close connection between magnetic moment and angular momentum. Both the magnetic moment and the angular momentum increase with the rate of rotation. The ratio of the two is called the gyromagnetic ratio and is simply the half of the charge-to-mass ratio.

For a spinning charged solid with a uniform charge density to mass density ratio, the gyromagnetic ratio is equal to half the charge-to-mass ratio. This implies that a more massive assembly of charges spinning with the same angular momentum will have a proportionately weaker magnetic moment, compared to its lighter counterpart. Even though atomic particles cannot be accurately described as spinning charge distributions of uniform charge-to-mass ratio, this general trend can be observed in the atomic world, where the intrinsic angular momentum (spin) of each type of particle is a constant: a small half-integer times the reduced Planck constant $\hbar$. This is the basis for defining the magnetic moment units of Bohr magneton (assuming charge-to-mass ratio of the electron) and nuclear magneton (assuming charge-to-mass ratio of the proton).

Effects of an External Magnetic Field on a Magnetic Moment

Force on a Moment

A magnetic moment in an externally produced magnetic field has a potential energy U:

$$U = -\mu \cdot B.$$

In a case when the external magnetic field is non-uniform, there will be a force, proportional to the magnetic field gradient, acting on the magnetic moment itself. There has been some discussion on how to calculate the force acting on a magnetic dipole. There are two expressions for the force acting on a magnetic dipole, depending on whether the model used for the dipole is a current loop or two monopoles (analogous to the electric dipole). The force obtained in the case of a current loop model is,

$$\mathbf{F}_{loop} = \nabla(\mu \cdot \mathbf{B}).$$

In the case of a pair of monopoles being used (i.e. electric dipole model):

$$\mathbf{F}_{dipole} = (\mu \cdot \nabla)\mathbf{B}.$$

and one can be put in terms of the other via the relation:

$$\mathbf{F}_{loop} = \mathbf{F}_{dipole} + \mu \times (\nabla \times \mathbf{B}).$$

In all these expressions μ is the dipole and B is the magnetic field at its position. Note that if there are no currents or time-varying electrical fields $\nabla \times B = 0$ and the two expressions agree.

An electron, nucleus, or atom placed in a uniform magnetic field will precess with a frequency known as the Larmor frequency.

Magnetic Dipoles

A magnetic dipole is the limit of either a current loop or a pair of poles as the dimensions of the source are reduced to zero while keeping the moment constant. As long as these limits only apply to fields far from the sources, they are equivalent. However, the two models give different predictions for the internal field.

External Magnetic Field Produced by a Magnetic Dipole Moment

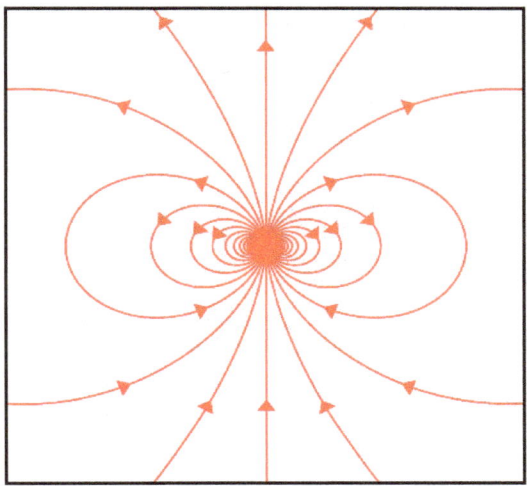

Magnetic field lines around a "magnetostatic dipole". The magnetic dipole itself is located in the center of the figure, seen from the side, and pointing upward.

Any system possessing a net magnetic dipole moment m will produce a dipolar magnetic field (described below) in the space surrounding the system. While the net magnetic field produced by the system can also have higher-order multipole components, those will drop off with distance more rapidly, so that only the dipolar component will dominate the magnetic field of the system at distances far away from it.

The vector potential of magnetic field produced by magnetic moment m is:

$$\mathbf{A}(\mathbf{r}) = \frac{\mu_0}{4\pi} \frac{\mathbf{m} \times \mathbf{r}}{|\mathbf{r}|^3}$$

and magnetic flux density is:

$$\mathbf{B}(\mathbf{r}) = \nabla \times \mathbf{A} = \frac{\mu_0}{4\pi} \left(\frac{3\mathbf{r}(\mathbf{m} \cdot \mathbf{r})}{|\mathbf{r}|^5} - \frac{\mathbf{m}}{|\mathbf{r}|^3} \right).$$

Alternatively one can obtain the scalar potential first from the magnetic pole perspective,

$$\psi(\mathbf{r}) = \frac{\mathbf{m} \cdot \mathbf{r}}{4\pi |\mathbf{r}|^3},$$

and hence magnetic field strength is:

$$\mathbf{H}(\mathbf{r}) = -\nabla \psi = \frac{1}{4\pi} \left(\frac{3\mathbf{r}(\mathbf{m} \cdot \mathbf{r})}{|\mathbf{r}|^5} - \frac{\mathbf{m}}{|\mathbf{r}|^3} \right).$$

The magnetic field of an ideal magnetic dipole is depicted on the right.

Internal Magnetic Field of a Dipole

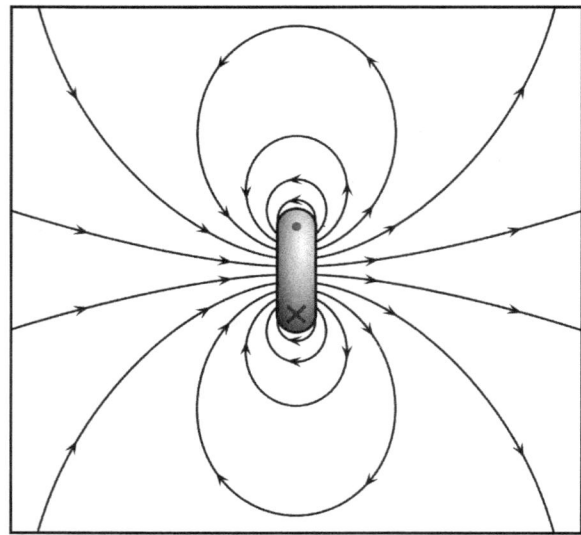

The magnetic field of a current loop.

The two models for a dipole (current loop and magnetic poles) give the same predictions for the magnetic field far from the source. However, inside the source region they give different predictions. The magnetic field between poles is in the opposite direction to the magnetic moment (which points from the negative charge to the positive charge), while inside a current loop it is in the same direction. Clearly, the limits of these fields must also be different as the sources shrink to zero size. This distinction only matters if the dipole limit is used to calculate fields inside a magnetic material.

If a magnetic dipole is formed by making a current loop smaller and smaller, but keeping the product of current and area constant, the limiting field is:

$$\mathbf{B}(\mathbf{x}) = \frac{\mu_0}{4\pi} \left[\frac{3\mathbf{n}(\mathbf{n} \cdot \mathbf{m}) - \mathbf{m}}{|\mathbf{x}|^3} + \frac{8\pi}{3} \mathbf{m} \delta(\mathbf{x}) \right].$$

This limit is correct for the internal field of the dipole.

If a magnetic dipole is formed by taking a "north pole" and a "south pole", bringing them closer and closer together but keeping the product of magnetic pole-charge and distance constant, the limiting field is,

$$H(x) = \frac{1}{4\pi}\left[\frac{3n(n\cdot m)-m}{|x|^3} - \frac{4\pi}{3}m\delta(x)\right].$$

These fields are related by $B = \mu_0(H + M)$, where $M(x) = m\delta(x)$ is the magnetization.

Forces between Two Magnetic Dipoles

As discussed earlier, the force exerted by a dipole loop with moment m_1 on another with moment m_2 is:

$$F = \nabla\left(m_2 \cdot B_1\right),$$

where B_1 is the magnetic field due to moment m_1. The result of calculating the gradient is,

$$F(r,m_1,m_2) = \frac{3\mu_0}{4\pi|r|^4}\left(m_2(m_1\cdot\hat{r}) + m_1(m_2\cdot\hat{r}) + \hat{r}(m_1\cdot m_2) - 5\hat{r}(m_1\cdot\hat{r})(m_2\cdot\hat{r})\right),$$

where $\hat{r}$ is the unit vector pointing from magnet 1 to magnet 2 and r is the distance. An equivalent expression is,

$$F = \frac{3\mu_0}{4\pi|r|^4}\left((\hat{r}\times m_1)\times m_2 + (\hat{r}\times m_2)\times m_1 - 2\hat{r}(m_1\cdot m_2) + 5\hat{r}(\hat{r}\times m_1)\cdot(\hat{r}\times m_2)\right).$$

The force acting on m_1 is in the opposite direction.

The torque of magnet 1 on magnet 2 is,

$$\tau = m_2 \times B_1.$$

Examples of Magnetic Moments

Two Kinds of Magnetic Sources

Fundamentally, contributions to any system's magnetic moment may come from sources of two kinds: motion of electric charges, such as electric currents; and the intrinsic magnetism of elementary particles, such as the electron.

Contributions due to the sources of the first kind can be calculated from knowing the distribution of all the electric currents (or, alternatively, of all the electric charges and their velocities) inside the system, by using the formulas below. On the other hand, the magnitude of each elementary particle's intrinsic magnetic moment is a fixed number, often measured experimentally to a great precision. For example, any electron's magnetic moment is measured to be -9.284764×10^{-24} J/T. The direction of the magnetic moment of any elementary particle is entirely determined by the direction of its spin, with the negative value indicating that any electron's magnetic moment is antiparallel to its spin.

The net magnetic moment of any system is a vector sum of contributions from one or both types of sources. For example, the magnetic moment of an atom of hydrogen-1 (the lightest hydrogen isotope, consisting of a proton and an electron) is a vector sum of the following contributions:

1. the intrinsic moment of the electron,

2. the orbital motion of the electron around the proton,

3. the intrinsic moment of the proton.

Similarly, the magnetic moment of a bar magnet is the sum of the contributing magnetic moments, which include the intrinsic and orbital magnetic moments of the unpaired electrons of the magnet's material and the nuclear magnetic moments.

Magnetic Moment of an atom

For an atom, individual electron spins are added to get a total spin, and individual orbital angular momenta are added to get a total orbital angular momentum. These two then are added using angular momentum coupling to get a total angular momentum. For an atom with no nuclear magnetic moment, the magnitude of the atomic dipole moment is then,

$$m_{Atom} = g_J \mu_B \sqrt{j(j+1)}.$$

where j is the total angular momentum quantum number, g_J is the Landé g-factor, and μ_B is the Bohr magneton. The component of this magnetic moment along the direction of the magnetic field is then,

$$m_{Atom}(z) = -m g_J \mu_B.$$

where m is called the magnetic quantum number or the *equatorial* quantum number, which can take on any of $2j + 1$ values:

$$-j, -(j-1) \cdots \dot{0} \cdots + (j-1), +j.$$

The negative sign occurs because electrons have negative charge.

Due to the angular momentum, the dynamics of a magnetic dipole in a magnetic field differs from that of an electric dipole in an electric field. The field does exert a torque on the magnetic dipole tending to align it with the field. However, torque is proportional to rate of change of angular momentum, so precession occurs: the direction of spin changes. This behavior is described by the Landau–Lifshitz–Gilbert equation:

$$\frac{1}{\gamma}\frac{d\mathbf{m}}{dt} = \mathbf{m} \times \mathbf{H}_{eff} - \frac{\lambda}{\gamma m}\mathbf{m} \times \frac{d\mathbf{m}}{dt},$$

where γ is the gyromagnetic ratio, m is the magnetic moment, λ is the damping coefficient and H_{eff} is the effective magnetic field (the external field plus any self-induced field). The first term describes precession of the moment about the effective field, while the second is a damping term related to dissipation of energy caused by interaction with the surroundings.

Magnetic Moment of an Electron

Electrons and many elementary particles also have intrinsic magnetic moments, an explanation of which requires a quantum mechanical treatment and relates to the intrinsic angular momentum of the particles as discussed in the article Electron magnetic moment. It is these intrinsic magnetic moments that give rise to the macroscopic effects of magnetism, and other phenomena, such as electron paramagnetic resonance.

The magnetic moment of the electron is:

$$\mathbf{m}_S = -\frac{g_S\mu_B\mathbf{S}}{\hbar},$$

where μ_B is the Bohr magneton, S is electron spin, and the g-factor g_S is 2 according to Dirac's theory, but due to quantum electrodynamic effects it is slightly larger in reality: 2.00231930436. The deviation from 2 is known as the anomalous magnetic dipole moment.

Again it is important to notice that m is a negative constant multiplied by the spin, so the magnetic moment of the electron is antiparallel to the spin. This can be understood with the following classical picture: if we imagine that the spin angular momentum is created by the electron mass spinning around some axis, the electric current that this rotation creates circulates in the opposite direction, because of the negative charge of the electron; such current loops produce a magnetic moment which is antiparallel to the spin. Hence, for a positron (the anti-particle of the electron) the magnetic moment is parallel to its spin.

Magnetic Moment of a Nucleus

The nuclear system is a complex physical system consisting of nucleons, i.e., protons and neutrons. The quantum mechanical properties of the nucleons include the spin among others. Since the electromagnetic moments of the nucleus depend on the spin of the individual nucleons, one can look at these properties with measurements of nuclear moments, and more specifically the nuclear magnetic dipole moment.

Most common nuclei exist in their ground state, although nuclei of some isotopes have long-lived excited states. Each energy state of a nucleus of a given isotope is characterized by a well-defined magnetic dipole moment, the magnitude of which is a fixed number, often measured experimentally to a great precision. This number is very sensitive to the individual contributions from nucleons, and a measurement or prediction of its value can reveal important information about the content of the nuclear wave function. There are several theoretical models that predict the value of the magnetic dipole moment and a number of experimental techniques aiming to carry out measurements in nuclei along the nuclear chart.

Magnetic Moment of a Molecule

Any molecule has a well-defined magnitude of magnetic moment, which may depend on the molecule's energy state. Typically, the overall magnetic moment of a molecule is a combination of the following contributions, in the order of their typical strength:

- Magnetic moments due to its unpaired electron spins (paramagnetic contribution), if any.

- Orbital motion of its electrons, which in the ground state is often proportional to the external magnetic field (diamagnetic contribution).

- The combined magnetic moment of its nuclear spins, which depends on the nuclear spin configuration.

Examples of Molecular Magnetism

- The dioxygen molecule, O_2, exhibits strong paramagnetism, due to unpaired spins of its outermost two electrons.

- The carbon dioxide molecule, CO_2, mostly exhibits diamagnetism, a much weaker magnetic moment of the electron orbitals that is proportional to the external magnetic field. The nuclear magnetism of a magnetic isotope such as ^{13}C or ^{17}O will contribute to the molecule's magnetic moment.

- The dihydrogen molecule, H_2, in a weak (or zero) magnetic field exhibits nuclear magnetism, and can be in a para- or an ortho- nuclear spin configuration.

- Many transition metal complexes are magnetic. The spin-only formula is a good first approximation for high-spin complexes of first-row transition metals.

Number of unpaired electrons	Spin-only moment (μ_B)
1	1.73
2	2.83
3	3.87
4	4.90
5	5.92

Elementary Particles

In atomic and nuclear physics, the Greek symbol μ represents the magnitude of the magnetic moment, often measured in Bohr magnetons or nuclear magnetons, associated with the intrinsic spin of the particle and/or with the orbital motion of the particle in a system. Values of the intrinsic magnetic moments of some particles are given in the table below:

Intrinsic magnetic moments and spins of some elementary particles		
Particle	Magnetic dipole moment (10^{-27} J·T^{-1})	Spin quantum number (dimensionless)
electron (e⁻)	−9284.764	1/2
proton (H⁺)	14.106067	1/2
neutron (n)	−9.66236	1/2
muon (μ⁻)	−44.904478	1/2
deuteron (²H⁺)	4.3307346	1
triton (³H⁺)	15.046094	1/2
helion (³He²⁺)	−10.746174	1/2
alpha particle (⁴He²⁺)	0	0

Magnetism in materials is crudely explained as mutual attraction between two pieces of a material, say iron or iron ore. There are various microscopic mechanisms of magnetism in materials. The strength of magnetism is quantitatively judged by a quantity called as 'magnetic moment'.

The major contributors of magnetic moment in a material are:

- Motion of electrons in an orbit of an atom. Orbital moment can be related to the current flowing in a loop of a wire of zero (negligible) resistance.

- Spinning of its electron around it own spin axis gives rise to a moment.

- Nuclear magnetic moment due to nuclei.

The first two contributions are quite significant and contribute to most of the magnetic character of a material while the third component, nuclear magnetic moment, is rather insignificant in the context of most magnetic materials of practical interest and can be neglected.

Magnetization

In classical electromagnetism, magnetization or magnetic polarization is the vector field that expresses the density of permanent or induced magnetic dipole moments in a magnetic material. The origin of the magnetic moments responsible for magnetization can be either microscopic electric currents resulting from the motion of electrons in atoms, or the spin of the electrons or the nuclei. Net magnetization results from the response of a material to an external magnetic field, together with any unbalanced magnetic dipole moments that may be inherent in the material itself; for example, in ferromagnets. Magnetization is not always uniform within a body, but rather varies between different points. Magnetization also describes how a material responds to an applied magnetic field as well as the way the material changes the magnetic field, and can be used to calculate the forces that result from those interactions. It can be compared to electric polarization, which is the measure of the corresponding response of a material to an electric field in electrostatics. Physicists and engineers usually define magnetization as the quantity of magnetic moment per unit volume. It is represented by a pseudovector M.

Definition

The magnetization field or M-field can be defined according to the following equation:

$$\mathbf{M} = \frac{d\mathbf{m}}{dV},$$

Where $d\mathbf{m}$ is the elementary magnetic moment and dV is the volume element; in other words, the M-field is the distribution of magnetic moments in the region or manifold concerned. This is better illustrated through the following relation:

$$\mathbf{m} = \iiint \mathbf{M}\,dV,$$

where m is an ordinary magnetic moment and the triple integral denotes integration over a volume. This makes the M-field completely analogous to the electric polarisation field, or P-field, used to determine the electric dipole moment p generated by a similar region or manifold with such a polarization:

$$\mathbf{P} = \frac{d\mathbf{p}}{dV}, \quad \mathbf{p} = \iiint \mathbf{P}\,dV,$$

Where $d\mathbf{p}$ is the elementary electric dipole moment.

Those definitions of P and M as a "moments per unit volume" are widely adopted, though in some cases they can lead to ambiguities and paradoxes.

The M-field is measured in *amperes per meter* (A/m) in SI units.

Physics Application

The magnetization is often not listed as a material parameter for commercially available ferromagnets. Instead the parameter that is listed is residual flux density, denoted $\mathbf{B_r}$. Physicists often need the magnetization to calculate the moment of a ferromagnet. To calculate the dipole moment m (A m²) using the formula:

$$\mathbf{m} = \mathbf{M}V,$$

we have that

$$\mathbf{M} = \mathbf{B_r} / \mu_0,$$

thus

$$\mathbf{m} = \mathbf{B_r}V / \mu_0,$$

where:

- Is the Residual Flux Density, expressed in Teslas (T).

- Is the volume (m³) of the magnet.

- H/m is the permeability of vacuum.

Magnetization in Maxwell's Equations

The behavior of magnetic fields (B, H), electric fields (E, D), charge density (ρ), and current density (J) is described by Maxwell's equations. The role of the magnetization is described below.

Relations between B, H, and M.

The magnetization defines the auxiliary magnetic field H as:

$$\mathbf{B} = \mu_0(\mathbf{H} + \mathbf{M}) \text{ (SI units)},$$

$$\mathbf{B} = \mathbf{H} + 4\pi\mathbf{M} \text{ (Gaussian units)},$$

which is convenient for various calculations. The vacuum permeability μ_0 is, by definition, $4\pi \times 10^{-7}$ V·s/(A·m).

A relation between M and H exists in many materials. In diamagnets and paramagnets, the relation is usually linear:

$$\mathbf{M} = \chi_m \mathbf{H},$$

where χ_m is called the volume magnetic susceptibility.

In ferromagnets there is no one-to-one correspondence between M and H because of Magnetic hysteresis.

Magnetization Current

The magnetization M makes a contribution to the current density J, known as the magnetization current.

$$\mathbf{J} = \nabla \times \mathbf{M},$$

and for the bound surface current:

$$\mathbf{K_m} = \mathbf{M} \times \hat{\mathbf{n}},$$

so that the total current density that enters Maxwell's equations is given by:

$$\mathbf{J} = \mathbf{J_f} + \nabla \times \mathbf{M} + \frac{\partial \mathbf{P}}{\partial t},$$

where J_f is the electric current density of free charges (also called the free current), the second term is the contribution from the magnetization, and the last term is related to the electric polarization P.

Magnetostatics

In the absence of free electric currents and time-dependent effects, Maxwell's equations describing the magnetic quantities reduce to:

$$\nabla \times \mathbf{H} = 0,$$
$$\nabla \cdot \mathbf{H} = -\nabla \cdot \mathbf{M},$$

These equations can be solved in analogy with electrostatic problems where:

$$\nabla \cdot \mathbf{E} = \frac{\rho}{\epsilon_0},$$
$$\nabla \times \mathbf{E} = 0,$$

In this sense $-\nabla \cdot M$ plays the role of a fictitious "magnetic charge density" analogous to the electric charge density ρ.

It is important to note that there is no such thing as a "magnetic charge," but that issue was still debated through the whole 19th century. Other concepts, that went along with it, such as the auxiliary field H, also have no real physical meaning in their own right. However, they are convenient mathematical tools, and are therefore still used today for applications such as modeling the magnetic field of the Earth.

Magnetization Dynamics

The time-dependent behavior of magnetization becomes important when considering nanoscale and nanosecond timescale magnetization. Rather than simply aligning with an applied field, the individual magnetic moments in a material begin to precess around the applied field and come into alignment through relaxation as energy is transferred into the lattice.

Reversal

Magnetization reversal, also known as switching, refers to the process that leads to a 180° (arc) re-orientation of the magnetization vector with respect to its initial direction, from one stable orientation to the opposite one. Technologically, this is one of the most important processes in magnetism that is linked to the magnetic data storage process such as used in modern hard disk drives. As it is known today, there are only a few possible ways to reverse the magnetization of a metallic magnet:

1. An applied magnetic field.

2. Spin injection via a beam of particles with spin.

3. Magnetization reversal by circularly polarized light; i.e., incident electromagnetic radiation that is circularly polarized.

Demagnetization

Demagnetization is the reduction or elimination of magnetization. One way to do this is to heat the object above its Curie temperature, where thermal fluctuations have enough energy to overcome exchange interactions, the source of ferromagnetic order, and destroy that order. Another way is to pull it out of an electric coil with alternating current running through it, giving rise to fields that oppose the magnetization.

One application of demagnetization is to eliminate unwanted magnetic fields. For example, magnetic fields can interfere with electronic devices such as cell phones or computers, and with machining by making cuttings cling to their parent.

Macroscopic View of Magnetization

We define magnetic induction, B, as $\mu_o H$ where μ_o is magnetic permeability of vacuum and its value is equal to $4\pi * 10^{-7} H.m^{-1}$. Units of B are tesla or Weber.m^{-2}.

This also explains that while H depends only on the current, B also depends upon the medium surrounding the wire which defines μ_o.

So now rewriting equation yields:

$$B = \mu_\circ NI,$$

Now imagine if a magnetic material is inserted within the coil, then there is a current induced in the magnetic material too, called Ameprian current, I_a which modifies the above equation to:

$$B = \mu_\circ NI + \mu_\circ I_a,$$

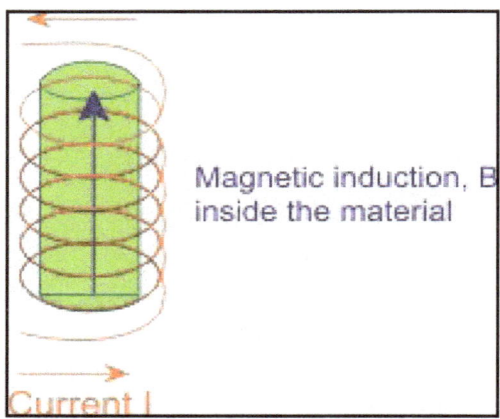

Magnetic induction, B inside the material

Current I

Magnetic Material inserted in the coil.

This Amperain current in the magnetic material can be replaced with induced magnetization, M, and hence using the above equation, we can write:

$$B = \mu_\circ (H + M),$$

If magnetization is assumed to be proportional to the magnetizing field, H, with proportionality constant defined as magnetic susceptibility, χ_m i.e. M = χ_m.H, we can write:

$$B = \mu_\circ (H + \chi_m M),$$

OR

$$B = \mu_\circ H(1 + \chi_m),$$

OR

$$B = \mu_\circ \mu_r H,$$

Here, χ_m, similar to dielectric materials, can be thought of as a parameter which expresses magnetic response of electron in a material to the applied magnetic field and is a dimensionless quantity. Here, $\mu_r = (1 + \chi_m)$ is the, in a similar manner to relative dielectric permittivity, ε_r.

In general, both susceptibility and permeability are tensors and assuming that vectors are collinear wherever there is vector notation is not used.

Naturally for vacuum, χ_m = 0.. However, unlike dielectric materials, χ_m can acquire both positive and negative values.

Classification of Magnetism

Magnetic materials can be classified based on the values of magnetic susceptibility.

Materials with negative susceptibility are diamagnetic ($\chi_m < 1$). Most diamagnetic materials show very small negative susceptibility except superconductors in superconducting effect when χ_m is equal to 1 which is very useful for applications such as magnetic levitation.

Materials with positive susceptibility are either paramagnetic, ferromagnetic or ferris-magnetic ($\chi_m > 1$). Susceptibility is positive but very small for paramagnetic materials but can be very large for ferro- and ferri-magnetic materials.

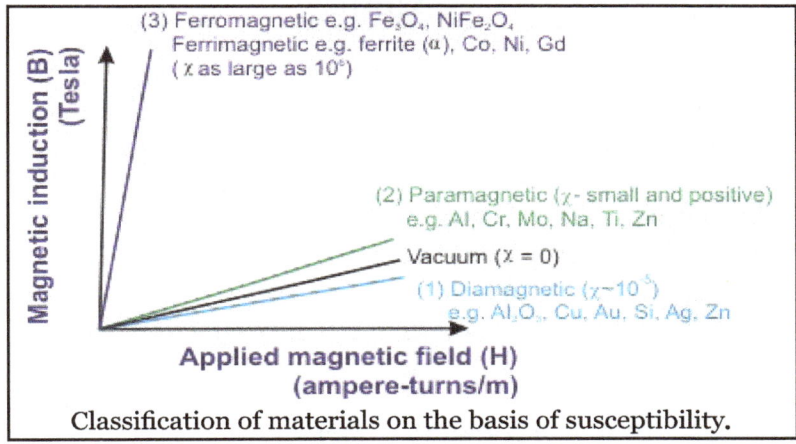

Classification of materials on the basis of susceptibility.

Susceptibility values of some of the common materials are provided below.

Material	χ(SI) unitless	χ(cgs) Unitless	μ Unitless	Type of Magnetism
Bi	-165×10^{-6}	-13.13×10^{-6}	0.99983	
Be	-23.2×10^{-6}	-1.85×10^{-6}	0.99998	
Ag	-23.2×10^{-6}	-1.90×10^{-6}	0.99997	
Au	-34.4×10^{-6}	-2.74×10^{-6}	0.99996	
Ge	-34.4×10^{-6}	-5.66×10^{-6}	0.99999	Diamagnetic
Cu	-9.7×10^{-6}	-0.77×10^{-6}	0.99999	
Si	-4.1×10^{-6}	-0.32×10^{-6}	0.99999	
Water	-9.14×10^{-6}	-0.73×10^{-6}	0.99999	
Superconductors (only in superconducting state)	-1.0	$\sim -8 \times 10^{-2}$	0	

β-Sn	$+2.4 \times 10^{-6}$	$+0.19 \times 10^{-6}$	1	
W	$+77.7 \times 10^{-6}$	$+6.18 \times 10^{-6}$	1.00008	**Paramagnetic**
Al	$+20.7 \times 10^{-6}$	$+1.65 \times 10^{-6}$	1.00002	
Pt	$+264.4 \times 10^{-6}$	$+21.04 \times 10^{-6}$	1.000026	
Low carbon steel	$\approx 5 \times 10^{3}$	3.98×10^{2}	5×10^{3}	
Fe-3%Si (Grain Oriented)	4×10^{3}	3.18×10^{3}	4×10^{4}	**Ferromagnetic**
Ni-Fe-Mo superalloy	10^{6}	7.96×10^{4}	10^{6}	

It should be noted, also as we will see, that except for diamagnetic materials, magnetic susceptibilities are temperature dependent. Sign of susceptibility can also be related (qualitatively) to the penetration of magnetic flux inside the material.

This says that for diamagnetic materials, when an external field is applied, the magnetic moment that is induced is in opposite direction to the field direction i.e. opposite magnetization as shown below. This is an inherent effect present in all materials. It is just that some materials like silver, gold, silicon are only diamagnetic i.e. they don't have any other effect present in them.

In many other materials, on top of the diamagnetic effect which is inherent to all materials, other effects are present which contribute significantly to the magnetization and all of these tend to have induced magnetization that is in the direction of the applied i.e. positive susceptibility. This means that magnetic flux penetrates into the material as shown below. We will discuss these effects one by one.

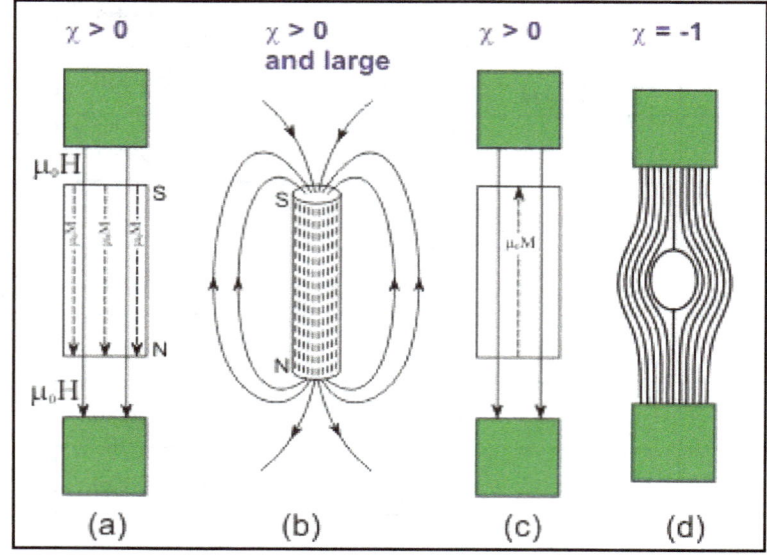

Schematic diagram showing flux penetration in magnetic materials with various ranges of susceptibilities.

In a nutshell, the net susceptibility of a material is the sum of all types of magnetic contributions. For materials having only diamagnetic contribution, this turns out to be negative.

Moreover, it is quantum mechanics which shows that materials are diamagnetic or paramagnetic or ferromagnetic. If taken too far, classical mechanics simply shows that all the magnetic moments in any material cancel out.

Diamagnetism

Diamagnetic materials are repelled by a magnetic field; an applied magnetic field creates an induced magnetic field in them in the opposite direction, causing a repulsive force. In contrast, paramagnetic and ferromagnetic materials are attracted by a magnetic field. Diamagnetism is a quantum mechanical effect that occurs in all materials; when it is the only contribution to the magnetism the material is called diamagnetic. In paramagnetic and ferromagnetic substances the weak diamagnetic force is overcome by the attractive force of magnetic dipoles in the material. The magnetic permeability of diamagnetic materials is less than μ_o, the permeability of vacuum. In most materials diamagnetism is a weak effect which can only be detected by sensitive laboratory instruments, but a superconductor acts as a strong diamagnet because it repels a magnetic field entirely from its interior.

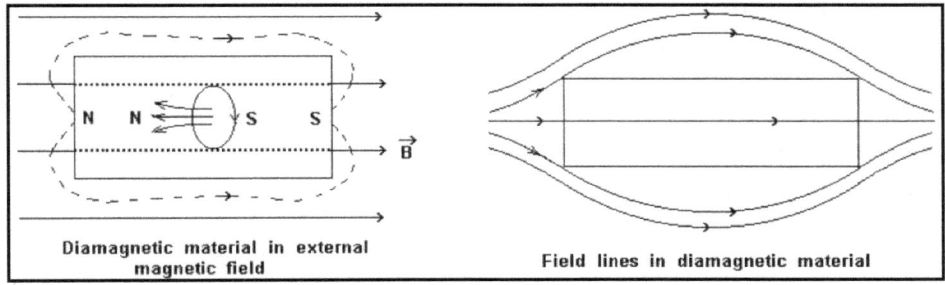

Dimagnetic material interaction in magnetic field.

Diamagnetism was first discovered when Sebald Justinus Brugmans observed in 1778 that bismuth and antimony were repelled by magnetic fields. In 1845, Michael Faraday demonstrated that it was a property of matter and concluded that every material responded (in either a diamagnetic or paramagnetic way) to an applied magnetic field. He adopted the term *diamagnetism* after it was suggested to him by William Whewell.

Materials

Notable diamagnetic materials	
Material	**χ_v ($\times 10^{-5}$)**
Superconductor	-10^5
Pyrolytic carbon	-40.9
Bismuth	-16.6
Mercury	-2.9

Notable diamagnetic materials	
Material	χ_v (× 10⁻⁵)
Silver	−2.6
Carbon (diamond)	−2.1
Lead	−1.8
Carbon (graphite)	−1.6
Copper	−1.0
Water	−0.91

Diamagnetism, to a greater or lesser degree, is a property of all materials and always makes a weak contribution to the material's response to a magnetic field. For materials that show some other form of magnetism (such as ferromagnetism or paramagnetism), the diamagnetic contribution becomes negligible. Substances that mostly display diamagnetic behaviour are termed diamagnetic materials, or diamagnets. Materials called diamagnetic are those that laymen generally think of as *non-magnetic*, and include water, wood, most organic compounds such as petroleum and some plastics, and many metals including copper, particularly the heavy ones with many core electrons, such as mercury, gold and bismuth. The magnetic susceptibility values of various molecular fragments are called Pascal's constants.

Diamagnetic materials, like water, or water-based materials, have a relative magnetic permeability that is less than or equal to 1, and therefore a magnetic susceptibility less than or equal to 0, since susceptibility is defined as $\chi_v = \mu_v - 1$. This means that diamagnetic materials are repelled by magnetic fields. However, since diamagnetism is such a weak property, its effects are not observable in everyday life. For example, the magnetic susceptibility of diamagnets such as water is $\chi_v = -9.05 \times 10^{-6}$. The most strongly diamagnetic material is bismuth, $\chi_v = -1.66 \times 10^{-4}$, although pyrolytic carbon may have a susceptibility of $\chi_v = -4.00 \times 10^{-4}$ in one plane. Nevertheless, these values are orders of magnitude smaller than the magnetism exhibited by paramagnets and ferromagnets. Note that because χ_v is derived from the ratio of the internal magnetic field to the applied field, it is a dimensionless value.

All conductors exhibit an effective diamagnetism when they experience a changing magnetic field. The Lorentz force on electrons causes them to circulate around forming eddy currents. The eddy currents then produce an induced magnetic field opposite the applied field, resisting the conductor's motion.

Superconductors

Superconductors may be considered perfect diamagnets ($\chi_v = -1$), because they expel all fields (except in a thin surface layer) due to the Meissner effect.

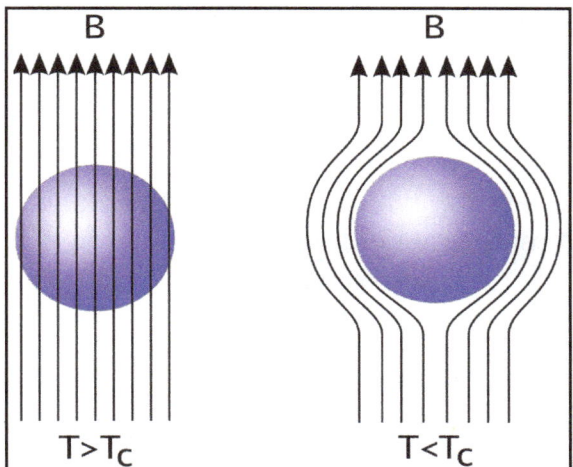

Transition from ordinary conductivity (left) to superconductivity (right). At the transition, the superconductor expels the magnetic field and then acts as a perfect diamagnet.

Demonstrations

Curving Water Surfaces

If a powerful magnet (such as a supermagnet) is covered with a layer of water (that is thin compared to the diameter of the magnet) then the field of the magnet significantly repels the water. This causes a slight dimple in the water's surface that may be seen by its reflection.

Levitation

A live frog levitates inside a 32 mm (1.26 in) diameter vertical bore of a Bitter solenoid in a magnetic field of about 16 teslas at the Nijmegen High Field Magnet Laboratory.

Diamagnets may be levitated in stable equilibrium in a magnetic field, with no power consumption. Earnshaw's theorem seems to preclude the possibility of static magnetic levitation. However, Earnshaw's theorem applies only to objects with positive susceptibilities, such as ferromagnets (which have a permanent positive moment) and paramagnets (which induce a positive moment). These are attracted to field maxima, which do not exist in free space. Diamagnets (which induce a negative moment) are attracted to field minima, and there can be a field minimum in free space.

A thin slice of pyrolytic graphite, which is an unusually strong diamagnetic material, can be stably floated in a magnetic field, such as that from rare earth permanent magnets. This can be done with all components at room temperature, making a visually effective demonstration of diamagnetism.

The Radboud University Nijmegen, the Netherlands, has conducted experiments where water and other substances were successfully levitated. Most spectacularly, a live frog was levitated.

In September 2009, NASA's Jet Propulsion Laboratory in Pasadena, California announced it had successfully levitated mice using a superconducting magnet, an important step forward since mice are closer biologically to humans than frogs. JPL said it hopes to perform experiments regarding the effects of microgravity on bone and muscle mass.

Recent experiments studying the growth of protein crystals have led to a technique using powerful magnets to allow growth in ways that counteract Earth's gravity.

A simple homemade device for demonstration can be constructed out of bismuth plates and a few permanent magnets that levitate a permanent magnet.

Theory

The electrons in a material generally circulate in orbitals, with effectively zero resistance and act like current loops. Thus it might be imagined that diamagnetism effects in general would be very, very common, since any applied magnetic field would generate currents in these loops that would oppose the change, in a similar way to superconductors, which are essentially perfect diamagnets. However, since the electrons are rigidly held in orbitals by the charge of the protons and are further constrained by the Pauli exclusion principle, many materials exhibit diamagnetism, but typically respond very little to the applied field.

The Bohr–van Leeuwen theorem proves that there cannot be any diamagnetism or paramagnetism in a purely classical system. However, the classical theory for Langevin diamagnetism gives the same prediction as the quantum theory. The classical theory is given below.

Langevin Diamagnetism

The Langevin theory of diamagnetism applies to materials containing atoms with closed shells. A field with intensity B, applied to an electron with charge e and mass

m, gives rise to Larmor precession with frequency $\omega = eB\,/\,2m$. The number of revolutions per unit time is $\omega\,/\,2\pi$, so the current for an atom with Z electrons is (in SI units):

$$I = -\frac{Ze^2 B}{4\pi m}.$$

The magnetic moment of a current loop is equal to the current times the area of the loop. Suppose the field is aligned with the z axis. The average loop area can be given as $\pi\langle\rho^2\rangle$, where $\langle\rho^2\rangle$ is the mean square distance of the electrons perpendicular to the z axis. The magnetic moment is therefore:

$$\mu = -\frac{Ze^2 B}{4m}\langle\rho^2\rangle.$$

If the distribution of charge is spherically symmetric, we can suppose that the distribution of x,y,z coordinates are independent and identically distributed. Then $\langle x^2\rangle = \langle y^2\rangle = \langle z^2\rangle = \frac{1}{3}\langle r^2\rangle$, where $\langle r^2\rangle$ is the mean square distance of the electrons from the nucleus. Therefore, $\langle\rho^2\rangle = \langle x^2\rangle + \langle y^2\rangle = \frac{2}{3}\langle r^2\rangle$. If N is the number of atoms per unit volume, the diamagnetic susceptibility in SI units is:

$$\chi = \frac{\mu_0 N\mu}{B} = -\frac{\mu_0 N Z e^2}{6m}\langle r^2\rangle.$$

In Metals

The Langevin theory does not apply to metals because they have non-localized electrons. The theory for the diamagnetism of a free electron gas is called Landau diamagnetism, and instead considers the weak counter-acting field that forms when their trajectories are curved due to the Lorentz force. Landau diamagnetism, however, should be contrasted with Pauli paramagnetism, an effect associated with the polarization of delocalized electrons' spins.

Diamagnetic materials are those in which the electron motions are such that they produce net zero magnetic moment in the absence of any magnetic field. Typically these are atoms with closed or filled outer electron shells.

Examples of such materials are inert gases, hydrogen, many metals (e.g. Ag, Au, Cu etc.), most non-metals (e.g. Si) and many organic compounds such as polymers.

Imagine a circular orbit of radius, r, around an atom with its center coinciding that of the atom. Now, we turn on the magnetic field in its vicinity.

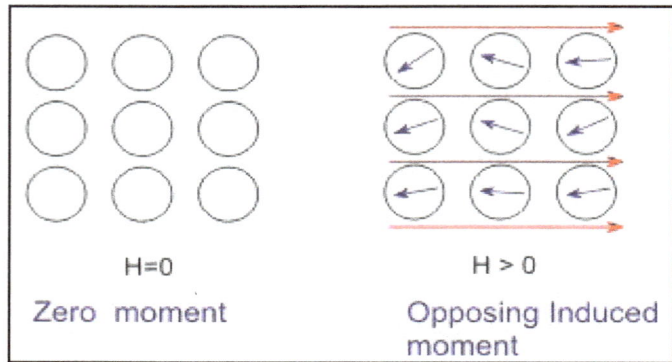

Representation of a diamagnetic material in the absence of a magnetic field and when a field is applied, note than when field is applied, induced moments oppose the field.

Hence, according to Faraday's law, as the magnetic field changes, it generates an electric field by magnetic induction. The electric field, E, tangent to the circular path is given as:

$$E.2\pi r = -\frac{d(B.\pi r^2)}{dt}$$

OR

$$E = -\frac{r}{2}\frac{dB}{dt}$$

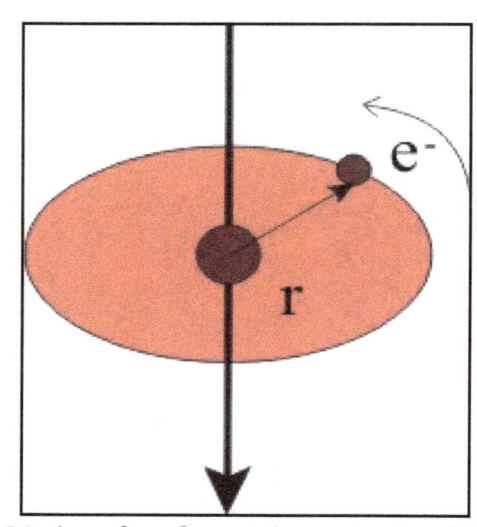

Motion of an electron in an atom's orbit.

This electric field produces a torque equivalent to -eE.r (i.e. F.r) which must be equal to the rate of change of angular momentum, J, i.e:

$$\frac{dJ}{dt} = e.\frac{r}{2}\frac{dB}{dt}.r$$

Two minuses cancel each other. Now integrating with respect to time with zero field, we get:

$$\Delta J = \left(\frac{er^2}{2}\right)B$$

This expression represents the extra angular momentum provided to the electrons when the field is applied.

Now, since the motion of electron is taking place in the orbits, the change in magnetic moment ($\Delta\mu_m$) which is orbital in nature is given as

Replacing $B = \mu_o H$, we get

$$\Delta\mu_m = -\left(\frac{e^2 r^2 \mu_0}{4m}\right) H,$$

For atoms with spherical symmetry, $\langle r^2 \rangle = (2/3) . \langle \bar{r}^2 \rangle$, hence,

$$\Delta\mu_m = -\frac{e^2 \bar{r}^2 \mu_0}{6m} H,$$

So, here we have a magnetic moment which is negative in sign to the magnetic field strength because it opposes the applied field. This magnetic moment is the moment for one electron.

So, if there are N electrons per unit volume, then magnetization, M, is given as:

$$M = -\frac{Ne^2 \bar{r}^2 \mu_2}{6m} H,$$

Hence magnetic susceptibility χ_m is given as:

$$X_m = \frac{M}{H} = -\frac{Ne^2 \bar{r}^2 \mu_0}{6m},$$

This equation shows the opposite nature of the magnetic susceptibility of the diamagnetic behaviour. Remember that diamagnetism is something which is present in all materials except that many materials also have other effects which completely overshadow the diamagnetic effect.

Paramagnetism

Paramagnetism is a form of magnetism whereby certain materials are weakly attracted by an externally applied magnetic field, and form internal, induced magnetic fields in the direction of the applied magnetic field. In contrast with this behavior, diamagnetic materials are repelled by magnetic fields and form induced magnetic fields in the direction opposite to that of the applied magnetic field. Paramagnetic materials include most chemical elements and some compounds; they have a relative magnetic permeability greater than or equal to 1 (i.e., a small positive magnetic susceptibility) and hence are attracted to magnetic fields. The magnetic moment induced by the applied field is linear in the field strength and rather weak. It typically requires a sensitive analytical balance to detect the effect and modern measurements on paramagnetic materials are often conducted with a SQUID magnetometer.

Paramagnetism is due to the presence of unpaired electrons in the material, so all atoms with incompletely filled atomic orbitals are paramagnetic. Due to their spin, unpaired electrons have a magnetic dipole moment and act like tiny magnets. An external magnetic field causes the electrons' spins to align parallel to the field, causing a net attraction. Paramagnetic materials include aluminum, oxygen, titanium, and iron oxide (FeO).

Unlike ferromagnets, paramagnets do not retain any magnetization in the absence of an externally applied magnetic field because thermal motion randomizes the spin orientations. (Some paramagnetic materials retain spin disorder even at absolute zero, meaning they are paramagnetic in the ground state, i.e. in the absence of thermal motion.) Thus the total magnetization drops to zero when the applied field is removed. Even in the presence of the field there is only a small induced magnetization because only a small fraction of the spins will be oriented by the field. This fraction is proportional to the field strength and this explains the linear dependency. The attraction experienced by ferromagnetic materials is non-linear and much stronger, so that it is easily observed, for instance, in the attraction between a refrigerator magnet and the iron of the refrigerator itself.

Relation to Electron Spins

Constituent atoms or molecules of paramagnetic materials have permanent magnetic moments (dipoles), even in the absence of an applied field. The permanent moment generally is due to the spin of unpaired electrons in atomic or molecular electron orbitals. In pure paramagnetism, the dipoles do not interact with one another and are randomly oriented in the absence of an external field due to thermal agitation, resulting in zero net magnetic moment. When a magnetic field is applied, the dipoles will tend to align with the applied field, resulting in a net magnetic moment in the direction of the applied field. In the classical description, this alignment can be understood to occur due to a torque being provided on the magnetic moments by an applied field, which tries to align the dipoles parallel to the applied field. However, the true origins of the alignment can only be understood via the quantum-mechanical properties of spin and angular momentum.

If there is sufficient energy exchange between neighbouring dipoles they will interact, and may spontaneously align or anti-align and form magnetic domains, resulting in ferromagnetism (permanent magnets) or antiferromagnetism, respectively. Paramagnetic behavior can also be observed in ferromagnetic materials that are above their Curie temperature, and in antiferromagnets above their Néel temperature. At these temperatures, the available thermal energy simply overcomes the interaction energy between the spins.

In general, paramagnetic effects are quite small: the magnetic susceptibility is of the order of 10^{-3} to 10^{-5} for most paramagnets, but may be as high as 10^{-1} for synthetic paramagnets such as ferrofluids.

Delocalization

Selected Pauli-paramagnetic metals	
Material	**Magnetic susceptibility, $\chi_v \left[10^{-5}\right]$**
Tungsten	6.8
Cesium	5.1
Aluminium	2.2
Lithium	1.4
Magnesium	1.2
Sodium	0.72

In conductive materials the electrons are delocalized, that is, they travel through the solid more or less as free electrons. Conductivity can be understood in a band structure picture as arising from the incomplete filling of energy bands. In an ordinary nonmagnetic conductor the conduction band is identical for both spin-up and spin-down electrons. When a magnetic field is applied, the conduction band splits apart into a spin-up and a spin-down band due to the difference in magnetic potential energy for spin-up and spin-down electrons. Since the Fermi level must be identical for both bands, this means that there will be a small surplus of the type of spin in the band that moved downwards. This effect is a weak form of paramagnetism known as *Pauli paramagnetism*.

The effect always competes with a diamagnetic response of opposite sign due to all the core electrons of the atoms. Stronger forms of magnetism usually require localized rather than itinerant electrons. However, in some cases a band structure can result in which there are two delocalized sub-bands with states of opposite spins that have different energies. If one subband is preferentially filled over the other, one can have itinerant ferromagnetic order. This situation usually only occurs in relatively narrow (d-)bands, which are poorly delocalized.

s and p Electrons

Generally, strong delocalization in a solid due to large overlap with neighboring wave functions means that there will be a large Fermi velocity; this means that the number of electrons in a band is less sensitive to shifts in that band's energy, implying a weak magnetism. This is why s- and p-type metals are typically either Pauli-paramagnetic or as in the case of gold even diamagnetic. In the latter case the diamagnetic contribution from the closed shell inner electrons simply wins from the weak paramagnetic term of the almost free electrons.

d and f Electrons

Stronger magnetic effects are typically only observed when d or f electrons are involved. Particularly the latter are usually strongly localized. Moreover, the size of the magnetic moment on a lanthanide atom can be quite large as it can carry up to 7 unpaired electrons in the case of gadolinium(III) (hence its use in MRI). The high magnetic moments associated with lanthanides is one reason why superstrong magnets are typically based on elements like neodymium or samarium.

Molecular Localization

Of course the above picture is a *generalization* as it pertains to materials with an extended lattice rather than a molecular structure. Molecular structure can also lead to localization of electrons. Although there are usually energetic reasons why a molecular structure results such that it does not exhibit partly filled orbitals (i.e. unpaired spins), some non-closed shell moieties do occur in nature. Molecular oxygen is a good example. Even in the frozen solid it contains di-radical molecules resulting in paramagnetic behavior. The unpaired spins reside in orbitals derived from oxygen p wave functions, but the overlap is limited to the one neighbor in the O_2 molecules. The distances to other oxygen atoms in the lattice remain too large to lead to delocalization and the magnetic moments remain unpaired.

Curie's Law

For low levels of magnetization, the magnetization of paramagnets follows what is known as Curie's law, at least approximately. This law indicates that the susceptibility, χ, of paramagnetic materials is inversely proportional to their temperature, i.e. that materials become more magnetic at lower temperatures. The mathematical expression is:

$$\mathbf{M} = \chi\mathbf{H} = \frac{C}{T}\mathbf{H}$$

where:

$\mathbf{M}$ is the resulting magnetization.

χ is the magnetic susceptibility.

$\mathbf{H}$ is the auxiliary magnetic field, measured in amperes/meter.

T is absolute temperature, measured in kelvins.

C is a material-specific Curie constant.

A derivation of this law:

Curie's law is valid under the commonly encountered conditions of low magnetization ($\mu_B H \lesssim k_B T$), but does not apply in the high-field/low-temperature regime where saturation of magnetization occurs ($\mu_B H \gtrsim k_B T$) and magnetic dipoles are all aligned with the applied field. When the dipoles are aligned, increasing the external field will not increase the total magnetization since there can be no further alignment.

For a paramagnetic ion with noninteracting magnetic moments with angular momentum J, the Curie constant is related the individual ions' magnetic moments,

$$C = \frac{N_A}{3k_B} \mu_{eff}^2 \ \ where \ \ \mu_{eff} = g_J \mu_B \sqrt{J(J+1)}$$

The parameter μ_{eff} is interpreted as the effective magnetic moment per paramagnetic ion. If one uses a classical treatment with molecular magnetic moments represented as discrete magnetic dipoles, μ, a Curie Law expression of the same form will emerge with μ appearing in place of μ_{eff}.

Curie's Law can be derived by considering a substance with noninteracting magnetic moments with angular momentum J. If orbital contributions to the magnetic moment are negligible (a common case), then in what follows J = S. If we apply a magnetic field along what we choose to call the z-axis, the energy levels of each paramagnetic center will experience Zeeman splitting of its energy levels, each with a z-component labeled by M_J (or just M_S for the spin-only magnetic case). Applying semiclassical Boltzmann statistics, the molar magnetization of such a substance is:

$$N_A \overline{m} = \frac{N_A \sum_{M_J=-J}^{J} \mu_{M_J} e^{-E_{M_J}/k_B T}}{\sum_{M_J=-J}^{J} e^{-E_{M_J}/k_B T}} = \frac{N_A \sum_{M_J=-J}^{J} M_J g_J \mu_B e^{M_J g_J \mu_B H/k_B T}}{\sum_{M_J=-J}^{J} e^{M_J g_J \mu_B H/k_B T}}$$

Where μ_{M_J} is the z-component of the magnetic moment for each Zeeman level, so $\mu_{M_J} = M_J g_J \mu_B - \mu_B$ is called the Bohr magneton and g_J is the Landé g-factor, which reduces to the free-electron g-factor, g_S when J = S. (in this treatment, we assume that the x- and y-components of the magnetization, averaged over all molecules, cancel out because the field applied along the z-axis leave them randomly oriented.) The energy of each Zeeman level is $E_{M_J} = -M_J g_J \mu_B H$. For temperatures over a few K, $M_J g_J \mu_B H / k_B T \ll 1$, and we can apply the approximation:

$$e^{M_J g_J \mu_B H/k_B T} \simeq 1 + M_J g_J \mu_B H / k_B T$$

$$\overline{m} = \frac{\sum_{M_J=-J}^{J} M_J g_J \mu_B e^{M_J g_J \mu_B H/k_B T}}{\sum_{M_J=-J}^{J} e^{M_J g_J \mu_B H/k_B T}} \simeq g_J \mu_B \frac{\sum_{M_J=-J}^{J} M_J \left(1 + M_J g_J \mu_B H / k_B T\right)}{\sum_{M_J=-J}^{J} \left(1 + M_J g_J \mu_B H / k_B T\right)} = \frac{g_J^2 \mu_B^2 H}{k_B T} \frac{\sum_{-J}^{J} M_J^2}{\sum_{M_J=-J}^{J} (1)},$$

which yields:

$$\bar{m} = \frac{g_J^2 \mu_B^2 H}{3k_B T} J(J+1).$$ The molar bulk magnetization is then,

$$M = N_A \bar{m} = \frac{N_A}{3k_B T}\left[g_J^2 J(J+1)\mu_B^2 \right] H,$$

and the molar susceptibility is given by

$$\chi_m = \frac{\partial M}{\partial H} = \frac{N_A}{3k_B T}\mu_{eff}^2 \; ; \quad and \quad \mu_{eff} = g_J \sqrt{J(J+1)}\mu_B.$$

When orbital angular momentum contributions to the magnetic moment are small, as occurs for most organic radicals or for octahedral transition metal complexes with d^3 or high-spin d^5 configurations, the effective magnetic moment takes the form (g_e = 2.0023... $\approx$ 2),

$$\mu_{eff} \simeq 2\sqrt{S(S+1)}\mu_B = \sqrt{n(n+2)}\mu_B,$$

where n is the number of unpaired electrons. In other transition metal complexes this yields a useful, if somewhat cruder, estimate.

Examples of Paramagnets

Materials that are called "paramagnets" are most often those that exhibit, at least over an appreciable temperature range, magnetic susceptibilities that adhere to the Curie or Curie–Weiss laws. In principle any system that contains atoms, ions, or molecules with unpaired spins can be called a paramagnet, but the interactions between them need to be carefully considered.

Systems with Minimal Interactions

The narrowest definition would be: a system with unpaired spins that *do not interact* with each other. In this narrowest sense, the only pure paramagnet is a dilute gas of monatomic hydrogen atoms. Each atom has one non-interacting unpaired electron. Of course, the latter could be said about a gas of lithium atoms but these already possess two paired core electrons that produce a diamagnetic response of opposite sign. Strictly speaking Li is a mixed system therefore, although admittedly the diamagnetic component is weak and often neglected. In the case of heavier elements the diamagnetic contribution becomes more important and in the case of metallic gold it dominates the properties. Of course, the element hydrogen is virtually never called 'paramagnetic' because the monatomic gas is stable only at extremely high temperature; H atoms combine to form molecular H_2 and in so doing, the magnetic moments are lost (*quenched*), because of the spins pair. Hydrogen is therefore *diamagnetic* and the same holds true for many other elements. Although the electronic configuration of the individual atoms (and ions) of most elements contain unpaired spins, they are not necessarily paramagnetic, because at ambient temperature quenching is very much the rule rather than the exception. The quenching tendency is weakest for f-electrons because f (especially $4f$) orbitals are radially contracted and they overlap only weakly with orbitals on adjacent atoms. Consequently, the lanthanide elements with incompletely filled 4f-orbitals are paramagnetic or magnetically ordered.

μ_{eff} values for typical d³ and d⁵ transition metal complexes.	
Material	μ_{eff}/μ_B
$[Cr(NH_3)_6]Br_3$	3.77
$K_3[Cr(CN)_6]$	3.87
$K_3[MoCl_6]$	3.79
$K_4[V(CN)_6]$	3.78
$[Mn(NH_3)_6]Cl_2$	5.92
$(NH_4)_2[Mn(SO_4)_2]\cdot 6H_2O$	5.92
$NH_4[Fe(SO_4)_2]\cdot 12H_2O$	5.89

Thus, condensed phase paramagnets are only possible if the interactions of the spins that lead either to quenching or to ordering are kept at bay by structural isolation of the magnetic centers. There are two classes of materials for which this holds:

- Molecular materials with a (isolated) paramagnetic center.

  o Good examples are coordination complexes of d- or f-metals or proteins with such centers, e.g. myoglobin. In such materials the organic part of the molecule acts as an envelope shielding the spins from their neighbors.

  o Small molecules can be stable in radical form, oxygen O_2 is a good example. Such systems are quite rare because they tend to be rather reactive.

- Dilute systems.

  o Dissolving a paramagnetic species in a diamagnetic lattice at small concentrations, e.g. Nd^{3+} in $CaCl_2$ will separate the neodymium ions at large enough distances that they do not interact. Such systems are of prime importance for what can be considered the most sensitive method to study paramagnetic systems: EPR.

Systems with Interactions

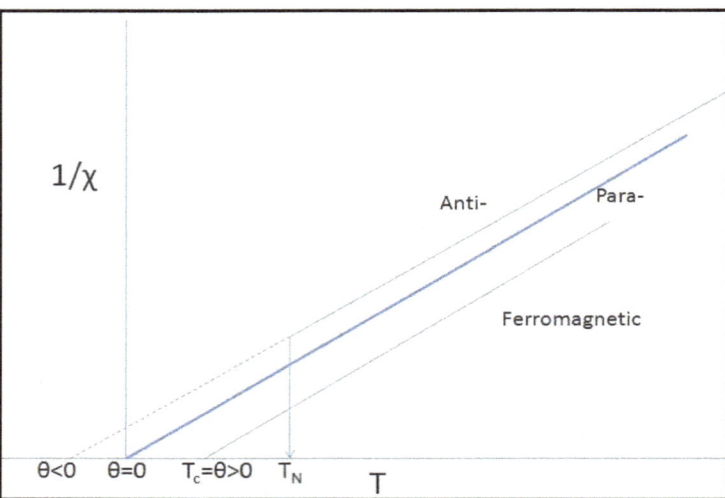

Idealized Curie–Weiss behavior; N.B. $T_C=\theta$, but T_N is not θ. Paramagnetic regimes are denoted by solid lines. Close to T_N or T_C the behavior usually deviates from ideal.

As stated above, many materials that contain d- or f-elements do retain unquenched spins. Salts of such elements often show paramagnetic behavior but at low enough temperatures the magnetic moments may order. It is not uncommon to call such materials 'paramagnets', when referring to their paramagnetic behavior above their Curie or Néel-points, particularly if such temperatures are very low or have never been properly measured. Even for iron it is not uncommon to say that *iron becomes a paramagnet* above its relatively high Curie-point. In that case the Curie-point is seen as a phase transition between a ferromagnet and a 'paramagnet'. The word paramagnet now merely refers to the linear response of the system to an applied field, the temperature dependence of which requires an amended version of Curie's law, known as the Curie–Weiss law:

$$\mathbf{M} = \frac{C}{T-\theta}\mathbf{H},$$

This amended law includes a term θ that describes the exchange interaction that is present albeit overcome by thermal motion. The sign of θ depends on whether ferro- or antiferromagnetic interactions dominate and it is seldom exactly zero, except in the dilute, isolated cases mentioned above.

Obviously, the paramagnetic Curie–Weiss description above T_N or T_C is a rather different interpretation of the word "paramagnet" as it does *not* imply the *absence* of interactions, but rather that the magnetic structure is random in the absence of an external field at these sufficiently high temperatures. Even if θ is close to zero this does not mean that there are no interactions, just that the aligning ferro- and the anti-aligning antiferromagnetic ones cancel. An additional complication is that the interactions are often different in different directions of the crystalline lattice (anisotropy), leading to complicated magnetic structures once ordered.

Randomness of the structure also applies to the many metals that show a net paramagnetic response over a broad temperature range. They do not follow a Curie type law as function of temperature however, often they are more or less temperature independent. This type of behavior is of an itinerant nature and better called Pauli-paramagnetism, but it is not unusual to see, for example, the metal aluminium called a "paramagnet", even though interactions are strong enough to give this element very good electrical conductivity.

Superparamagnets

Some materials show induced magnetic behavior that follows a Curie type law but with exceptionally large values for the Curie constants. These materials are known as superparamagnets. They are characterized by a strong ferromagnetic or ferrimagnetic type of coupling into domains of a limited size that behave independently from one another. The bulk properties of such a system resembles that of a paramagnet, but on a microscopic level they are ordered. The materials do show an ordering temperature above which the behavior reverts to ordinary paramagnetism (with interaction). Ferrofluids are a good example, but the phenomenon can also occur inside solids, e.g., when dilute paramagnetic centers are introduced in a strong itinerant medium of ferromagnetic coupling such as when Fe is substituted in $TlCu_2Se_2$ or the alloy AuFe. Such systems contain ferromagnetically coupled clusters that freeze out at lower temperatures. They are also called mictomagnets.

In paramagnetic materials, atoms have a permanent non-zero net magnetic moment due to the sum of orbital and spin magnetic moments. However at room temperature, in paramagnetic materials, thermal energy causes random distribution of magnetic moments and hence net magnetization appears to be zero for the whole material (not just an atom).

Upon application of a field, the moments tend to align up in the direction of the field overcoming the thermal barrier and giving a net positive magnetic moment in the direction of the applied field. The susceptibilities of these materials are usually very small, 10^{-3} to 10^{-6}.

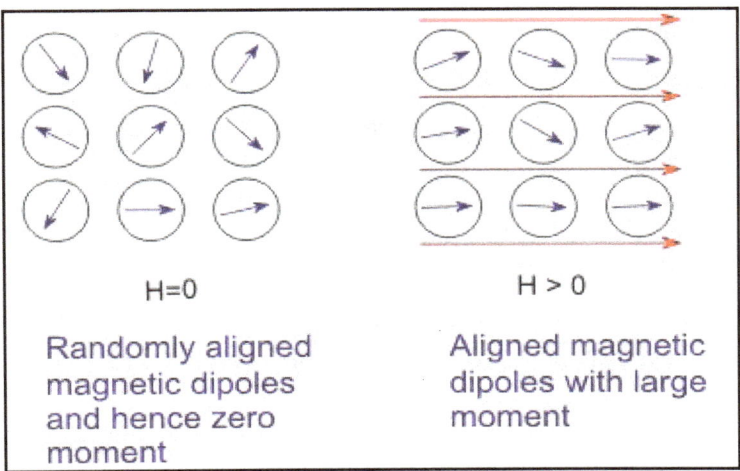

H=0

H > 0

Randomly aligned magnetic dipoles and hence zero moment

Aligned magnetic dipoles with large moment

Schematic representation of spins in a paramagnetic solid.

In most solids, only spin paramagnetism is observed, since the electron orbits are considered as coupled to the lattice (the orbital moments are quenched) and hence do not contribute significantly to the magnetic moment. Qualitatively, it is due to the electric field generated by the surrounding ions in the solid. Because of these electric fields, the orbitals are strongly coupled to the lattice and hence cannot reorient themselves along the field and therefore, do not contribute towards the magnetic moment. On the other hand, electron spins are weakly coupled and hence form a major part of the magnetic moment.

Spin paramagnetism is observed in materials with unfilled d-shells which follow Hund's rule which says that in the partially filled d-orbitals, the moment is maximized by the first filling all the spin states of one direction followed by filling in other direction.

For example, Ni, with an atomic number of 28 has an electronic configuration of $1s^2$, $2s^2$, $2p^6$, $3s^2$, $3p^6$, $4s^2$ and $3d^8$. The filling in of partially occupied d-orbital is achieved as follows:

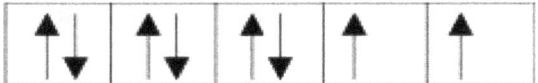

As a result, each Ni atom has a net magnetization of $2\mu_B$.

Exception to this may be rare-earth elements and their derivatives which have deep-lying 4f- electrons. These are shielded by the outer electrons from the crystal field and as result they show both spin and orbital paramagnetism.

Susceptibility (from orbital moment) in paramagnetic material shows temperature dependence as governed by Curie's law and is given as:

$$\chi_{para}^{orbit} = \frac{M}{H} = \frac{N\mu_m^2 \mu_\circ}{3kT} = \frac{C}{T}$$

where N is the number of atoms per unit volume, μ_m is the magnetic moment of an atom and C is Curie constant. Curie's law states that susceptibility of a paramegntic material is inversely proportional to temperature.

Here we assume that each atom has a magnetic moment μ_m whose magnitude is the same but the direction can be random. The magnetic energy in a field B would be $\mu_m.B = -\mu_m B \cos\alpha$ where α is the angle between the moment and field.

Boltzmann statistics in classical thermodynamics gives the probability of having any angle i.e. of occupying any energy as $\exp(-\mu_m H.\cos\alpha/kT)$. As a result, as you can also notice that it is more plausible to have moment closer to $0°$ than $180°$ with respect to the applied field, simply because that leads to smaller energy than those at $180°$.

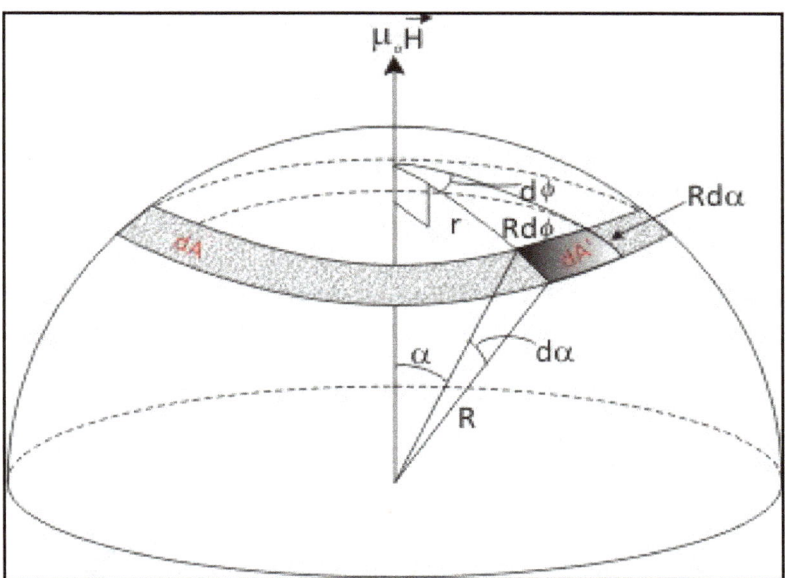

Representation of magnetic moment in a paramagnetic solid with respect to the applied field.

Consider a smaller number of dipoles dn making an angle a with respect to the applied field and penetrating an area dA as shown in figure which is proportional to exp(-μ_mH-cosα/kT)

Considering a unit radius of the sphere (R=1), the area dA can be worked out as:

$$dA = 2\pi r^2 \sin\alpha \, d\alpha.$$

Hence dn can be expressed as:

$$dn = C.dA.\exp\left(-\mu_m^{H\cos\alpha}/_{kT}\right)$$

or

$$dn = C.2\pi r^2 \sin\alpha \, d\alpha \exp\left(-\mu_m^{H\cos\alpha}/_{kT}\right)$$

where C is a constant.

Assuming that γ = μ_mH/kT, we can integrate the above equation for a between 0 to 180° which will yield the total number of dipoles i.e.

$$N = \int_0^\pi C.2\pi r^2 \sin\alpha \, d\alpha \exp(-\gamma\cos\alpha) \, d\alpha$$

Now, the total magnetic moment along the direction of magnetic field can be expressed multiplication of number dipoles multiplied by the magnetic moment of each dipole along the field direction ($\mu_m$cosα) and is given as:

$$M_{tot} = \int_0^\pi \mu_m \cos\alpha.C.2\pi r^2 \sin\alpha \, d\alpha \exp\left(-\mu_m^{H\cos\alpha}/_{kT}\right)$$

Dividing equations yields the average magnetization, M_{avj}, as:

$$M_{avj} = \frac{\int_0^\pi \mu_m \cos\alpha . C.2\pi r^2 \sin\alpha \, d\alpha \exp\left(-\mu_m{}^{H\cos\alpha}/_{kT}\right)}{\int_0^\pi C.2\pi r^2 \sin\alpha \, d\alpha \exp(-\gamma\cos\alpha) d\alpha}$$

$$= \frac{\mu_m \int_0^\pi \cos\alpha . C.2\pi r^2 \sin\alpha \, d\alpha \exp(-\gamma\cos\alpha) d\alpha}{\int_0^\pi \sin\alpha \, d\alpha \exp(-\gamma\cos\alpha) d\alpha}$$

Now, following orientation polarization, we can show that for reasonably small magnetic field, moments are parallel to applied field and magnetization M can be expressed as:

$$M = N\mu_m . \left\{ \coth\left(\frac{\mu_m B}{kT}\right) - \frac{kT}{\mu_m B} \right\} = N\mu_m L(\beta)$$

where $\beta = (\mu_m B)(kT)$ and $L(\beta) = \cot h\beta - 1/\beta$ is called Langevin function. For most cases i.e. when $\mu_m B << kT$, we can assume $L(\beta) = \beta/3$. Using this approximation, we can write equation (28) as:

$$M = \frac{N\mu_m{}^2 B}{3kT} = \frac{N\mu_m{}^2 \mu_\circ H}{3kT}$$

OR

$$\chi_{para} = \frac{M}{H} = \frac{C}{T}$$

where Curie constant $C = (N\mu_m{}^2 \mu_\circ)/3k$.

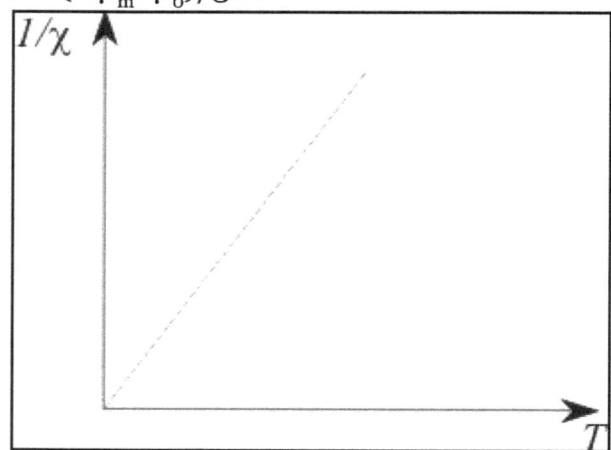

Susceptibility vs temperature plot for a paramagnetic material.

Equations show that induced magnetization is proportional to the applied field and is larger at lower temperature. The same result can also be shown using quantum mechanism which is not a part of this course.

Quantum mechanics, for any characteristic number j (could be combined for both orbital and spin moments or only *s* for spin moments), also gives:

$$\chi_{para} = g^2 j(j+1)\frac{N\mu_B^2\mu_\circ}{3kT} = \frac{C}{T},$$

which is similar to $\mu_m = \mu_{eff} = g.\sqrt{j(j+1)}.\mu_B$. Here, g is Landé-g factor.

The theoretically derived and experimentally obtained values of magnetic moment for a few selected magnetic ions are given below and here, you can see while for rare earth ions total magnetic moment value gives a much better estimate, for transition elements, spin only magnetism in a better estimate:

Ion	Configuration	Calculated		Measured
		$g.\sqrt{j(j+1)}$	$g.\sqrt{s(s+1)}$	$d(\mu_m/\mu_B)$
Rare Earths				
Ce³⁺	$4f^1\,5s^2\,5p^6$	2.54	-	2.4
Pr³⁺	$4f^2\,5s^2\,5p^6$	3.58	-	3.5
Nd³⁺	$4f^3\,5s^2\,5p^6$	3.62	-	3.5
Transition Elements				
Mn³⁺	$3d^4$	0.00	4.90	4.9
Mn²⁺, Fe³⁺	$3d^5$	5.92	5.92	5.9
Fe²⁺	$3d^6$	6.70	4.90	5.4
Co²⁺	$3d^7$	6.63	3.87	4.8
Ni²⁺	$3d^8$	5.59	2.83	3.2

Ferrite (Magnet)

A ferrite is a type of ceramic compound composed of iron(III) oxide (Fe_2O_3) combined chemically with one or more additional metallicelements. They are both electrically nonconductive and ferrimagnetic, meaning they can be magnetized or attracted to a magnet. Ferrites can be divided into two families based on their magnetic coercivity, their resistance to being demagnetized. *Hard ferrites* have high coercivity, hence they are difficult to demagnetize. They are used to make magnets, for devices such as refrigerator magnets, loudspeakers and small electric motors. *Soft ferrites* have low coercivity. They are used in the electronics industry to make ferrite cores for inductors and transformers, and in various microwave components. Ferrite compounds have extremely low cost, being made of iron oxide (i.e. rusted iron), and also have excellent corrosion resistance. They are very stable and difficult to demagnetize, and can be made with both high and low coercive forces. Yogoro Kato and Takeshi Takei of the Tokyo Institute of Technology synthesized the first ferrite compounds in 1930.

Composition and Properties

Ferrites are usually non-conductive ferrimagnetic ceramic compounds derived from iron oxides such as hematite (Fe_2O_3) or magnetite (Fe_3O_4) as well as oxides of other metals. Ferrites are, like most of the other ceramics, hard and brittle.

Many ferrites are spinels with the formula AB_2O_4, where A and B represent various metal cations, usually including iron Fe. Spinel ferrites usually adopt a crystal motif consisting of cubic close-packed (fcc) oxides (O^{2-}) with A cations occupying one eighth of the tetrahedral holes and B cations occupying half of the octahedral holes. If one eighth of the tetrahedral holes are taken by B cation, then one fourth of the octahedral sites are occupied by A cation and the other one fourth by B cation and it is called the inverse spinel structure. It is also possible to have mixed structure spinel ferrites with formula $[M^{2+}_{1-\delta}Fe^{3+}_{\delta}][M^{2+}_{\delta}Fe^{3+}_{2-\delta}]O_4$ where δ is the degree of inversion.

The magnetic material known as "ZnFe" has the formula $ZnFe_2O_4$, with Fe^{3+} occupying the octahedral sites and Zn^{2+} occupy the tetrahedral sites, it is an example of normal structure spinel ferrite.

Some ferrites have hexagonal crystal structure, like barium and strontium ferrites $BaFe_{12}O_{19}$ ($BaO:6Fe_2O_3$) and $SrFe_{12}O_{19}$ ($SrO:6Fe_2O_3$).

In terms of their magnetic properties, the different ferrites are often classified as "soft" or "hard", which refers to their low or high magnetic coercivity, as follows.

Soft Ferrites

Various ferrite cores used to make small transformers and inductors.

Ferrites that are used in transformer or electromagnetic cores contain nickel, zinc, and/ or manganese compounds. They have a low coercivity and are called soft ferrites. The low coercivity means the material's magnetization can easily reverse direction without dissipating much energy (hysteresis losses), while the material's high resistivity prevents eddy currents in the core, another source of energy loss. Because of their comparatively low losses at high frequencies, they are extensively used in the cores of RF transformers and inductors in applications such as switched-mode power supplies and loopstick antennas used in AM radios.

The most common soft ferrites are:

- Manganese-zinc ferrite (MnZn, with the formula $Mn_aZn_{(1-a)}Fe_2O_4$). MnZn have higher permeability and saturation induction than NiZn.

- Nickel-zinc ferrite (NiZn, with the formula $Ni_aZn_{(1-a)}Fe_2O_4$). NiZn ferrites exhibit higher resistivity than MnZn, and are therefore more suitable for frequencies above 1 MHz.

For applications below 5 MHz, MnZn ferrites are used; above that, NiZn is the usual choice. The exception is with common mode inductors, where the threshold of choice is at 70 MHz.

Hard Ferrites

In contrast, permanent ferrite magnets are made of hard ferrites, which have a high coercivity and high remanence after magnetization. Iron oxide and barium or strontium carbonate are used in manufacturing of hard ferrite magnets. The high coercivity means the materials are very resistant to becoming demagnetized, an essential characteristic for a permanent magnet. They also have high magnetic permeability. These so-called *ceramic magnets* are cheap, and are widely used in household products such as refrigerator magnets. The maximum magnetic field B is about 0.35 tesla and the magnetic field strength H is about 30 to 160 kiloampere turns per meter (400 to 2000 oersteds). The density of ferrite magnets is about 5 g/cm³.

The most common hard ferrites are:

- Strontium ferrite, $SrFe_{12}O_{19}$ ($SrO·6Fe_2O_3$), used in small electric motors, micro-wave devices, recording media, magneto-optic media, telecommunication and electronic industry.

- Barium ferrite, $BaFe_{12}O_{19}$ ($BaO·6Fe_2O_3$), a common material for permanent magnet applications. Barium ferrites are robust ceramics that are generally stable to moisture and corrosion-resistant. They are used in e.g. loudspeaker magnets and as a medium for magnetic recording, e.g. on magnetic stripe cards.

- Cobalt ferrite, $CoFe_2O_4$ ($CoO·Fe_2O_3$), used in some media for magnetic recording.

Production

Ferrites are produced by heating a mixture of finely-powdered precursors pressed into a mold. During the heating process, calcination of carbonates occurs:

$$MCO_3 \rightarrow MO + CO_2$$

The oxides of barium and strontium are typically supplied as their carbonates, $BaCO_3$ or $SrCO_3$. The resulting mixture of oxides undergoes sintering. Sintering is a high temperature process similar to the firing of ceramic ware.

Afterwards, the cooled product is milled to particles smaller than 2 µm, small enough that each particle consists of a single magnetic domain. Next the powder is pressed into a shape, dried, and re-sintered. The shaping may be performed in an external magnetic field, in order to achieve a preferred orientation of the particles (anisotropy).

Small and geometrically easy shapes may be produced with dry pressing. However, in such a process small particles may agglomerate and lead to poorer magnetic properties compared to the wet pressing process. Direct calcination and sintering without re-milling is possible as well but leads to poor magnetic properties.

Electromagnets are pre-sintered as well (pre-reaction), milled and pressed. However, the sintering takes place in a specific atmosphere, for instance one with an oxygen shortage. The chemical composition and especially the structure vary strongly between the precursor and the sintered product.

To allow efficient stacking of product in the furnace during sintering and prevent parts sticking together, many manufacturers separate ware using ceramic powder separator sheets. These sheets are available in various materials such as alumina, zirconia and magnesia. They are also available in fine, medium and coarse particle sizes. By matching the material and particle size to the ware being sintered, surface damage and contamination can be reduced while maximizing furnace loading.

Uses

Ferrite cores are used in electronic inductors, transformers, and electromagnets where the high electrical resistance of the ferrite leads to very low eddy current losses. They are commonly seen as a lump in a computer cable, called a ferrite bead, which helps to prevent high frequency electrical noise (radio frequency interference) from exiting or entering the equipment.

Early computer memories stored data in the residual magnetic fields of hard ferrite cores, which were assembled into arrays of *core memory*. Ferrite powders are used in the coatings of magnetic recording tapes. One such type of material is iron (III) oxide.

Ferrite particles are also used as a component of radar-absorbing materials or coatings used in stealth aircraft and in the absorption tiles lining the rooms used for electromagnetic compatibility measurements.

Most common radio magnets, including those used in loudspeakers, are ferrite magnets. Ferrite magnets have largely displaced Alnico magnets in these applications.

It is a common magnetic material for electromagnetic instrument pickups.

Ferrite nanoparticles exhibit superparamagnetic properties.

History

Yogoro Kato and Takeshi Takei of the Tokyo Institute of Technology synthesized the first ferrite compounds in 1930. This led to the founding of TDK Corporation in 1935, to manufacture the material.

Barium hexaferrite ($BaFe_{12}O_{19}$) was discovered in 1950 at the Philips Natuurkundig Laboratorium (*Philips Physics Laboratory*). The discovery was somewhat accidental—due to a mistake by an assistant who was supposed to be preparing a sample of hexagonal lanthanum ferrite for a team investigating its use as a semiconductor material. On discovering that it was actually a magnetic material, and confirming its structure by X-ray crystallography, they passed it on to the magnetic research group. Barium hexaferrite has both high coercivity (170 kA/m) and low raw material costs. It was developed as a product by Philips Industries (Netherlands) and from 1952 was marketed under the trade name *Ferroxdure*. The low price and good performance led to a rapid increase in the use permanent magnets.

In the 1960s Philips developed strontium hexaferrite ($SrFe_{12}O_{19}$), with better properties than barium hexaferrite. Barium and strontium hexaferrite dominate the market due to their low costs. However other materials have been found with improved properties. $BaFe^{2+}_2Fe^{3+}_{16}O_{27}$ came in 1980 and $Ba_2ZnFe_{18}O_{23}$ came in 1991.

Ferromagnetism

Basic Characteristics

In addition to permanent magnetic moments as contained in paramagnetic materials, ferromagnetic materials consist of ordered regions or domains of single orientation of magnetic moment giving rise to large finite magnetization in the absence of a magnetic field, much like polarization in ferroelectric materials. This phenomenon is observed below a critical temperature called as Curie Temperature, above which the material behave like a paramagnetic material.

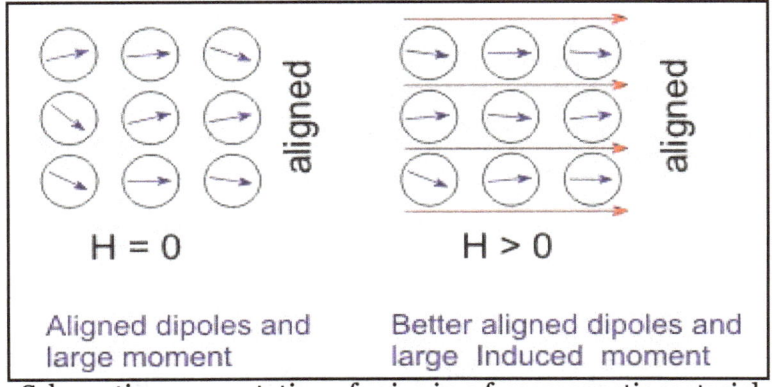

Schematic representation of spins in a ferromagnetic material.

When a varying magnetic field is applied to a ferromagnetic material, the material exhibits a ferroelectric-like hysteresis loop between magnetization and the magnetic field as shown below. In this figure, notice how the domain structure changes with field in the initial stages of magnetization.

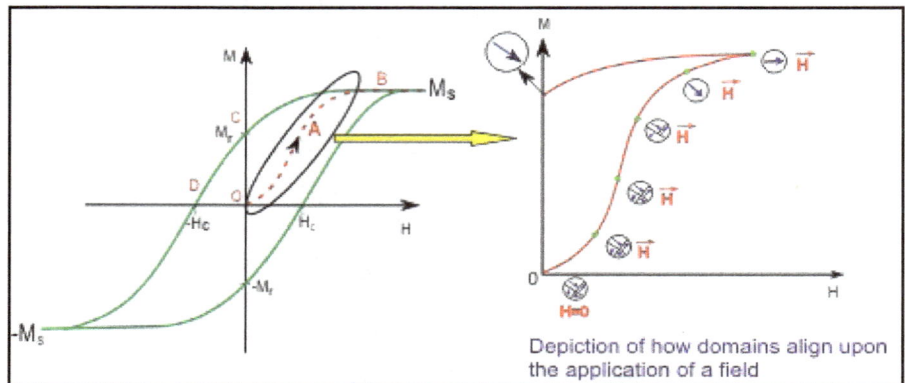

Ferromagnetic hysteresis loops.

Most of the ferromagnetic materials are elemental metals such as iron, nickel, cobalt etc. However some oxides such as chrmoimum oxide, CrO_2, are ferromagnetic oxides. These oxides also tend to be conducting and behave like metals.

Antiferromagnetic Materials

These are materials in which electron spins associated with magnetic atoms at particular crystallographic sites are ordered yet oriented with respect to each other in such a manner that their net magnetization is equal to zero. This is the case below a particular temperature, called as Néel temperature (T_N) above which the material behaves as a paramagnet.

Examples include metallic manganese, chromium, various transition metal oxides such as manganese oxide (MnO), forms of iron oxide (Fe_2O_3), multiferroic perovskites like bismuth ferrite ($BiFeO_3$). For example in MnO, since O is not a magnetic ion, the antiparallel spin arrangement of Mn^{2+}, the magnetic ion, in two sites gives rise to zero magnetization. Below is the crystal structure of MnO, drawn on (100) plane.

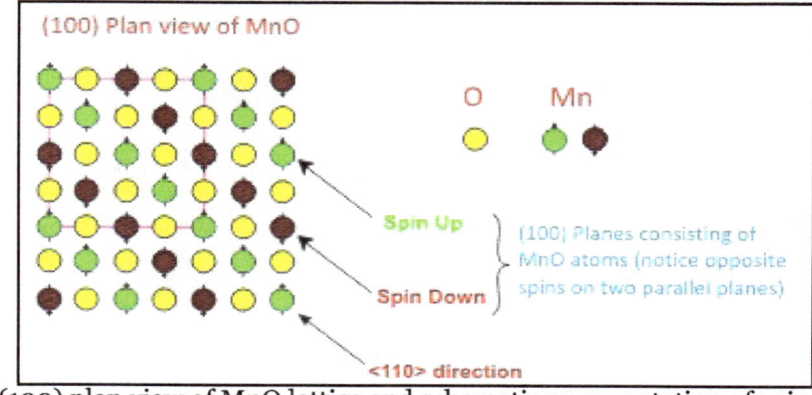

(100) plan view of MnO lattice and schematic representation of spins.

As shown below, the susceptibility of an antiferromagnetic material shows a paramagnetic (PM) behavior above (T_N) and between 0 K and (T_N), it shows an antiferromagnetic (AFM) behavior.

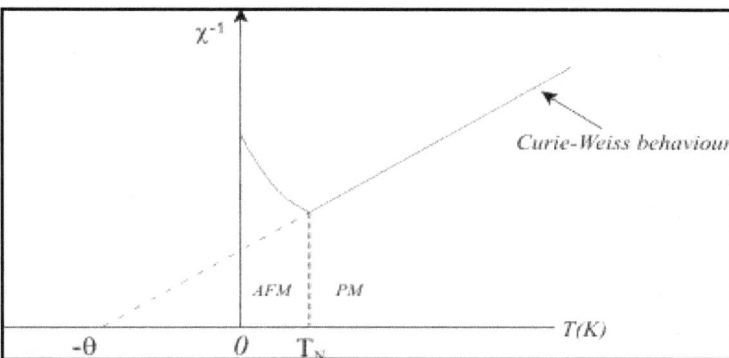

Temperature dependence of susceptibility of an antiferromagnetic material.

Antiferromagnetic materials are not very useful as such because they do not show any magnetization whatsoever.

Some of the materials like bismuth ferrite ($BiFeO_3$) have been described as canted antiferromagnets where spin of atoms are not exactly antiparallel but are slightly inclined towards the so-called parallel axis which may give rise to very small magnetization.

Antiferromagnetism

In materials that exhibit antiferromagnetism, the magnetic moments of atoms or molecules, usually related to the spins of electrons, align in a regular pattern with neighboring spins (on different sublattices) pointing in opposite directions. This is, like ferromagnetism and ferrimagnetism, a manifestation of ordered magnetism. Generally, antiferromagnetic order may exist at sufficiently low temperatures, vanishing at and above a certain temperature, the Néel temperature (named after Louis Néel, who had first identified this type of magnetic ordering). Above the Néel temperature, the material is typically paramagnetic.

Measurement

When no external field is applied, the antiferromagnetic structure corresponds to a vanishing total magnetization. In an external magnetic field, a kind of ferrimagnetic behavior may be displayed in the antiferromagnetic phase, with the absolute value of one of the sublattice magnetizations differing from that of the other sublattice, resulting in a nonzero net magnetization. Although the net magnetization should be zero at a temperature of absolute zero, the effect of spin canting often causes a small net magnetization to develop.

The magnetic susceptibility of an antiferromagnetic material typically shows a maximum at the Néel temperature. In contrast, at the transition between the ferromagnetic to the paramagnetic phases the susceptibility will diverge. In the antiferromagnetic case, a divergence is observed in the *staggered susceptibility*.

Various microscopic (exchange) interactions between the magnetic moments or spins may lead to antiferromagnetic structures. In the simplest case, one may consider an Ising model on a bipartite lattice, e.g. the simple cubic lattice, with couplings between spins at nearest neighbor sites. Depending on the sign of that interaction, ferromagnetic or antiferromagnetic order will result. Geometrical frustration or competing ferro- and antiferromagnetic interactions may lead to different and, perhaps, more complicated magnetic structures.

Antiferromagnetic materials occur commonly among transition metal compounds, especially oxides. Examples include hematite, metals such as chromium, alloys such as iron manganese (FeMn), and oxides such as nickel oxide (NiO). There are also numerous examples among high nuclearity metal clusters. Organic molecules can also exhibit antiferromagnetic coupling under rare circumstances, as seen in radicals such as 5-dehydro-m-xylylene.

Antiferromagnets can couple to ferromagnets, for instance, through a mechanism known as exchange bias, in which the ferromagnetic film is either grown upon the antiferromagnet or annealed in an aligning magnetic field, causing the surface atoms of the ferromagnet to align with the surface atoms of the antiferromagnet. This provides the ability to "pin" the orientation of a ferromagnetic film, which provides one of the main uses in so-called spin valves, which are the basis of magnetic sensors including modern hard drive read heads. The temperature at or above which an antiferromagnetic layer loses its ability to "pin" the magnetization direction of an adjacent ferromagnetic layer is called the blocking temperature of that layer and is usually lower than the Néel temperature.

Geometric Frustration

Unlike ferromagnetism, anti-ferromagnetic interactions can lead to multiple optimal states (ground states—states of minimal energy). In one dimension, the anti-ferromagnetic ground state is an alternating series of spins: up, down, up, down, etc. Yet in two dimensions, multiple ground states can occur.

Consider an equilateral triangle with three spins, one on each vertex. If each spin can take on only two values (up or down), there are $2^3 = 8$ possible states of the system, six of which are ground states. The two situations which are not ground states are when all three spins are up or are all down. In any of the other six states, there will be two favorable interactions and one unfavorable one. This illustrates frustration: the inability of the system to find a single ground state. This type of magnetic behavior has been found in minerals that have a crystal stacking structure such as a Kagome lattice or hexagonal lattice.

Other Properties

Synthetic antiferromagnets (often abbreviated by SAF) are artificial antiferromagnets consisting of two or more thin ferromagnetic layers separated by a nonmagnetic layer. Due to dipole coupling of the ferromagnetic layers results in antiparallel alignment of the magnetization of the ferromagnets.

Antiferromagnetism plays a crucial role in giant magnetoresistance, as had been discovered in 1988 by the Nobel prize winners Albert Fert and Peter Grünberg (awarded in 2007) using synthetic antiferromagnets.

There are also examples of disordered materials (such as iron phosphate glasses) that become antiferromagnetic below their Néel temperature. These disordered networks 'frustrate' the antiparallelism of adjacent spins; i.e. it is not possible to construct a network where each spin is surrounded by opposite neighbour spins. It can only be determined that the average correlation of neighbour spins is antiferromagnetic. This type of magnetism is sometimes called *speromagnetism*.

Ferrimagnetic Materials

These are materials which again, like antiferromagnetic materials, show antiparallel alignment of moments at particular atomic sites i.e. magnetic moment of one crystal sub-lattice is anti-parallel to the other. But since most of these materials consist of cations of two or more types, sub-lattices contain two different types of ions with different magnetic moment for two types of atoms and as a result, net magnetization is not equal to zero. The examples of such materials are various kinds of cubic spinel ferrites such as $NiFe_2O_4$, $CoFe_2O_4$, Fe_3O_4 (or $FeO.Fe_2O_3$), $CuFe_2O_4$ etc. Other examples are hexagonal ferrites likes $BaFe_{12}O_{19}$, garnets such as $Y_3Fe_5O_{12}$, represented by a general formula $R_3Fe_5O_{12}$ where R, in addition to yttrium can be one of lanthanide atoms such as lanthanum, cerium, samarium etc.

A schematic representation of this inequality in the neighbouring magnetic moment can be like this:

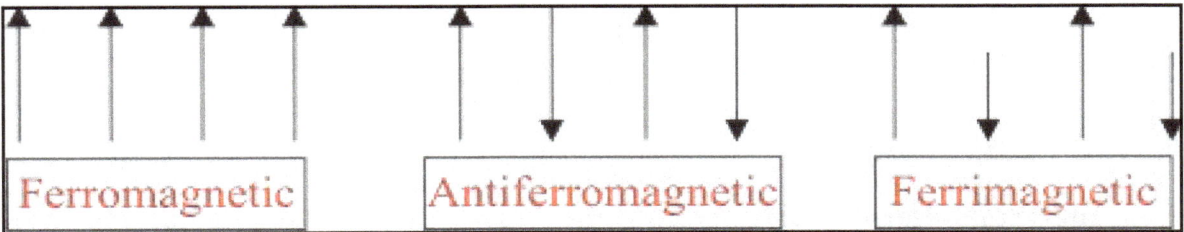

Magnetic moment arrangements in magnetically ordered materials.

These materials also follow a temperature dependence of magnetization and susceptibility near Curie transition (actually Néel transition) in a similar manner as shown by the ferromagnetic materials. These materials, like ferromagnetic materials, show significantly large magnetization below the magnetic transition temperature and hence, often the temperature dependent behavior is clubbed with that of ferromagnetic materials.

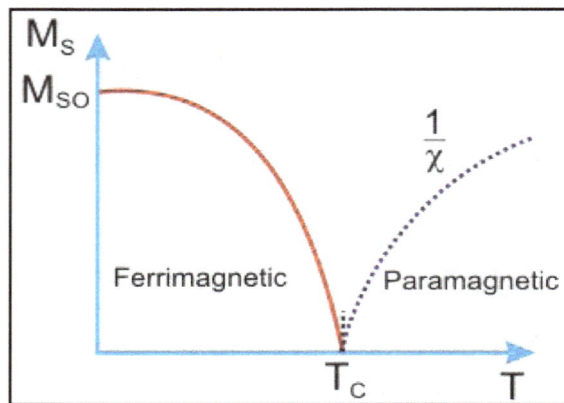

Temperature dependence of magnetization and susceptibility in a ferrimagnetic material.

In physics, a ferrimagnetic material is one that has populations of atoms with opposing magnetic moments, as in antiferromagnetism; however, in ferrimagnetic materials, the opposing moments are unequal and a spontaneous magnetization remains. This happens when the populations consist of different materials or ions (such as Fe^{2+} and Fe^{3+}).

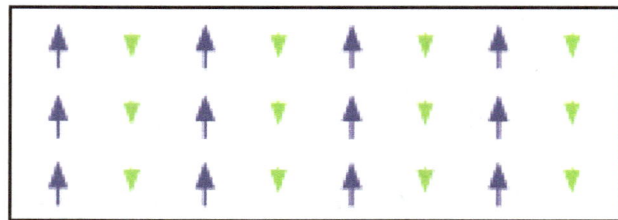

Ferrimagnetic ordering.

Ferrimagnetism is exhibited by ferrites and magnetic garnets. The oldest known magnetic substance, magnetite (iron(II,III) oxide; Fe_3O_4), is a ferrimagnet; it was originally classified as a ferromagnet before Néel's discovery of ferrimagnetism and antiferromagnetism in 1948.

Known ferrimagnetic materials include YIG (yttrium iron garnet), cubic ferrites composed of iron oxides and other elements such as aluminum, cobalt, nickel, manganese and zinc, hexagonal ferrites such as $PbFe_{12}O_{19}$ and $BaFe_{12}O_{19}$, and pyrrhotite, $Fe_{1-x}S$.

Effects of Temperature

Ferrimagnetic materials are like ferromagnets in that they hold a spontaneous magnetization below the Curie temperature, and show no magnetic order (are paramagnetic) above this temperature. However, there is sometimes a temperature *below* the Curie temperature at which the two opposing moments are equal, resulting in a net magnetic moment of zero; this is called the *magnetization compensation point*. This compensation point is observed easily in garnets and rare earth-transition metal alloys (RE-TM). Furthermore, ferrimagnets may also have an *angular momentum compensation point* at which the net angular momentum vanishes. This compensation point is a crucial point for achieving high speed magnetization reversal in magnetic memory devices.

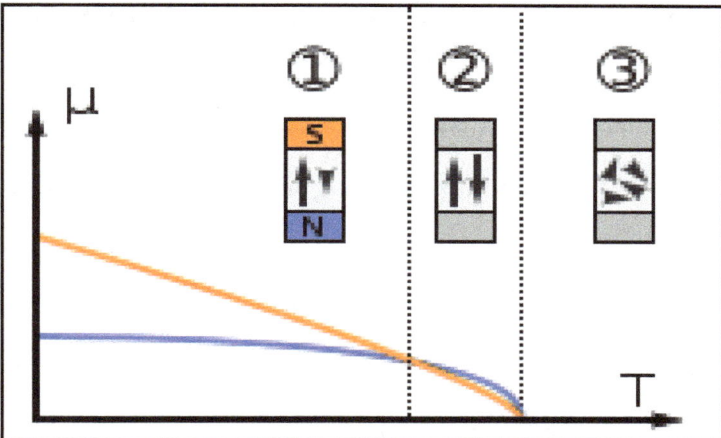

① Below the magnetization compensation point, ferrimagnetic material is magnetic. ② At the compensation point, the magnetic components cancel each other and the total magnetic moment is zero. ③ Above the Curie point, the material loses magnetism.

Properties

Ferrimagnetic materials have high resistivity and have anisotropic properties. The anisotropy is actually induced by an external applied field. When this applied field aligns with the magnetic dipoles it causes a net magnetic dipole moment and causes the magnetic dipoles to precess at a frequency controlled by the applied field, called *Larmor* or *precession frequency*. As a particular example, a microwave signal circularly polarized in the same direction as this precession strongly interacts with the magnetic dipole moments; when it is polarized in the opposite direction the interaction is very low. When the interaction is strong, the microwave signal can pass through the material. This directional property is used in the construction of microwave devices like isolators, circulators and gyrators. Ferrimagnetic materials are also used to produce optical isolators and circulators. Ferrimagnetic minerals in various rock types are used to study ancient geomagnetic properties of Earth and other planets. That field of study is known as paleomagnetism.

Molecular ferrimagnets

Ferrimagnetism can also occur in molecular magnets. A classic example is a dodecanuclear manganese molecule with an effective spin of S = 10 derived from antiferromagnetic interaction on Mn(IV) metal centres with Mn(III) and Mn(II) metal centres.

Comparison between Different Kinds of Magnetism

So far, we have learnt that there are four kinds of magnetism i.e. diamagnetism, paramagnetism, ferromagetism, and ferri- and antiferromagnetism. The field dependent magnetization behavior of these different types is shown below as well the dependence on temperature. Diamagnetism is a basic character present in every material, it is just that this behavior is overshadowed in materials where other effects are present too.

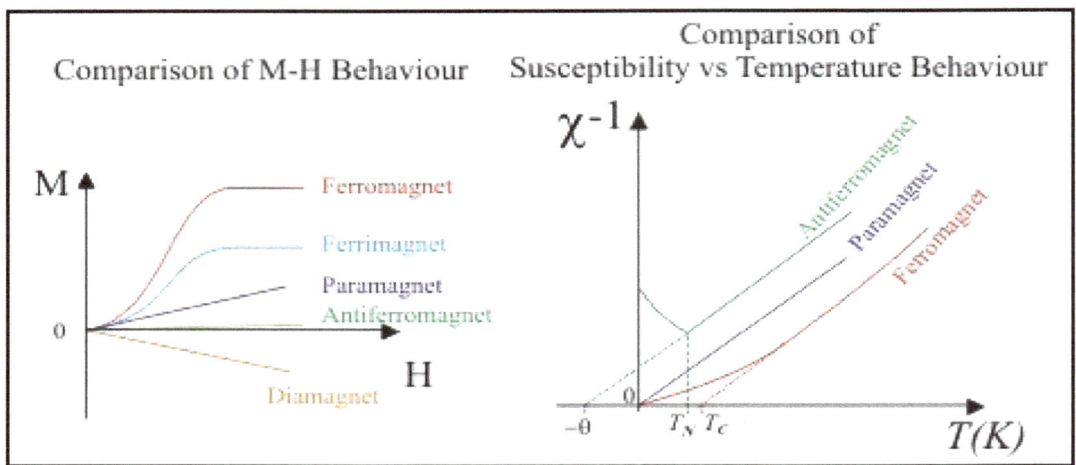

Comparison of magnetic behaviour (magnetization and susceptibility) of various kinds of magnetic materials versus temperature.

Practically most important materials are ferromagnets, mostly metals, and ferrimagentic materials, mostly ferrite oxides.

Magnetic Losses and Frequency Dependence

This is important mainly for ferromagnetic and ferrimagnetic materials when they are placed in an alternating magnetic field. Associated with frequency response are magnetic hysteresis losses and frequency response of magnetic permeability.

Magnetic Losses

We imagine a ferromagnetic or ferrimagnetic material with a hysteresis curve is subjected to an alternating magnetic field, $H = H_o \exp(i\omega t)$. The frequencies are kept such, typically low, such that switching of magnetization of the material keeps pace with the magnetic field switching. In an analogy to the dielectric materials, two loss mechanisms are of importance:

- Eddy current losses induced by the alternating magnetic field which are quite dependent on the resistivity of the material;

- The domain wall motion requires as well as dissipates some amount of energy resulting in what are called as intrinsic magnetic losses or hysteresis losses.

Both of these effects add up and energy loss is dissipated as heat. While frequency, magnetic field strength and maximum magnetic flux increase both types of losses, increasing the resistivity (ρ) decreases eddy current losses in particular, losses are more when the material is conducting and many ferromagnets tend to be quite conductive.

Hysteresis loss is the energy loss per cycle times number of cycles and is given as:

$$P_H = 2.f.H_c.B_r$$

where f is the frequency of applied field, H_c is the coercive field strength and B_r is magnetic remanence.

Eddy current power losses can be expressed as:

$$P_e = \frac{\pi d^2}{6\rho}(f.B_{max})$$

The total power loss is the sum of these two losses.

These two equations can help a user to choose a material carefully so that losses can be minimized.

Frequency Response of Permeability

It is given that ferro- and ferri-magnetic materials have permanent moments which constitute regions of aligned moment called as domains with two regions of different regions separated by a domain wall. Now, the ease of domain wall movement can be described in terms of coercivity, i.e. the higher the coercive field, the easier it is for domains walls to move during switching.

So, all we can say at this point is that this domain wall movement is dependent on the frequency of the applied field. In some materials, low frequencies are needed for easy switching while some materials can witness easy switching at GHz or microwave region frequencies making them suitable for microwave devices.

This also means that permeability, μ, in a similar manner as permittivity in dielectrics, can be defined as a complex quantity i.e. $\mu^* = \mu' + \mu''$ where μ' is the real part of permeability and μ'' is the complex part of the permeability.

Categories of Magnetic Ferrites

Cubic Spinel Ferrites

Cubic spinel ferrites have a formula AB_2O_4 which crystallize with a face centered cubic structure. In these structures, two cations occupy tetrahedral and octahedral sites in an FCC lattice made by O atoms. One unit-cell consists of eight formula units of AB_2O_4 hence containing a total of 32 octahedral interstices with one fourth occupancy and 64 tetrahedral interstices with one eighth occupancy by the cations.

Depending on how these cations are distributed in the interstices, cubic spinel structures can be of two types:

- Normal spinel,

- Inverse spinel.

From the magnetism point of interest, the cations occupying tetrahedral sites have their spins oppositely oriented with respect to the cations on octahedral sites (up and down depends upon your frame of reference). Compounds with normal spinel structure are $ZnFe_2O_4$, $MgAl_2O_4$, $CoAl_2O_4$ where A atoms occupy the tetrahedral sites while B atoms occupy the octahedral sites.

In compounds with inverse spinel strcture e.g. Fe_3O_4, $NiFe_2O_4$, half of B cations occupy tetrahedral sites and all A and the remaining 50% B cations occupy octahedral sites.

So, if we take the example of Fe_3O_4 which is actually $Fe^{2+}Fe^{3+}_2O_4$, then the arrangement of spin is like what is shown below.

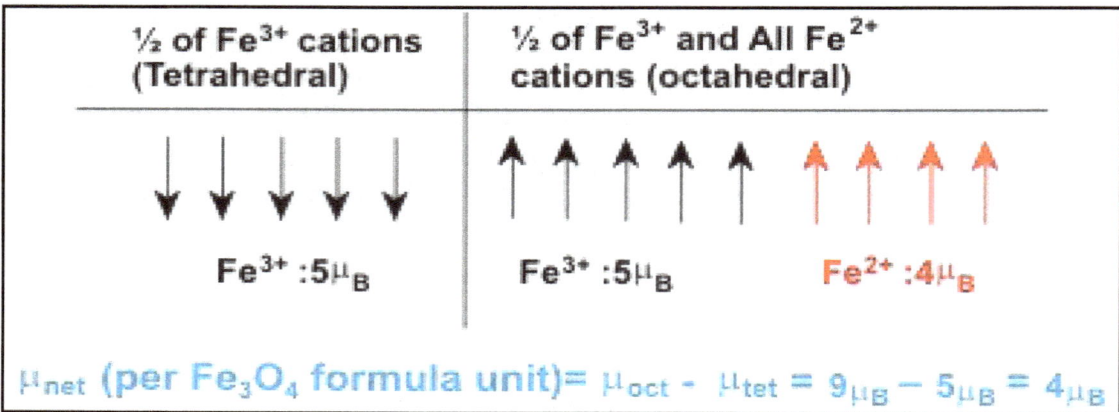

Hence the net magnetization of Fe_3O_4 is $4\mu_B$ per formula unit which is quite a large magnetic moment. You can convert this into $A.m^{-1}$ by simply calculating the net moment for the whole unit cell and dividing that by the cell volume.

Similarly you can work out magnetic moments of other ferrites such as $NiFe_2O_4$, $CoFe_2O_4$ etc. This approach is also valid for mixed spinel compounds.

Mixed spinels are quite a nice way of increasing the net magnetic moment. As you can see that since the moments of two sites are antiparallel, reducing the net magnetic moment of one site would actually increase the net moment. So mixing of $NiFe_2O_4$, an inverse spinel, and $ZnFe_2O_4$, a normal spinel, results in maximization of moment up to ~40 mol% $ZnFe_2O_4$.

In general, spinel ferrites show low magnetic anisotropy i.e. dependence of magnetization on crystallographic directions, and are magnetically soft i.e. show low coercive fields. Exceptions could be Co-containing ferrites which are not only strongly magnetically anisotropic but also show large coercive fields strengths. These materials also exhibit a ferromagnetic material like hysteresis loop when placed in a varying magnetic field.

Hexagonal Ferrites

Hexagonal ferrites are based on hexagonal magnetoplubite and are often called M-type ferrites. The model compound of this family is barium ferrite with formula $BaFe_{12}O_{19}$. The large hexagonal unit-cell contains 64 atoms, i.e. two formula units. The structure is basically a mixture of cubic closed packed and hexagonal closed packed layers formed by barium and oxygen ions. Chemical substitution of Ba sites is usually done with Sr atoms while Fe atoms are substituted by Al atoms, based on the size and valence, resulting in a change in the magnetic behavior.

Out of 12 iron atoms of one formula unit, 9 occupy the octahedral sites, two occupy tetrahedral sites and the remaining one is 5-fold coordinated. Out of these, 7 atoms on the octahedral site and 1 with 5-fold coordination have their spins in one direction while spins of the rest of the atoms are oriented oppositely i.e. say 8 atoms with spins up and 4 atoms with spin down. As we saw earlier, each Fe^{3+} ion has spin magnetic moment of $5\mu_B$ simple math gives a net magnetic moment of $20\mu_B$ per formula unit leading to a magnetic moment of $40\mu_B$ per unit cell.

This material has a high degree of magnetic anisotropy and it magnetizes relatively easily along -direction or c-axis of its unit cell. The material is typically categorized as a hard ferrite with coercivity between 50-100 $kA.m^1$ depending upon the microstructure and composition.

Garnets

Garnets are usually known as minerals. In the context of magnetic materials, garnets are represented by a general formula $Y_3Fe_5O_{12}$, containing two magnetic ions, one typically being iron and another being rare earth. Here R, in addition to yttrium can be one of lanthanide atoms such as lanthanum, cerium, samarium etc.

The unit cell of $Y_3Fe_5O_{12}$ is cubic and contains 8 formula units i.e. 160 atoms, quite complex! In garnet ferrites, orbital magnetic contribution of iron atoms is quenched due to shielding from crystal field while lanthanide ions contribute to both orbital and spin magnetic moment, thus contributing more to the total magnetic moment.

In this structure, R atoms are cubic coordinated i.e. 12-fold coordinated, 2 Fe atoms are octahedrally coordinated and the remaining three Fe atoms are tetrahedrally coordinated with antiparallel spin configuration of spins on tetrahedral and octahedral sites while orientation of spins on R-site is parallel to those on octahedral sites. We know that each Fe^{3+} ion contributes $5\mu_B$ which each lanthanide atom, R, contributes a moment of magnitude $\mu_R\mu_B$ where μ_R is the strength of moment of R ion. Hence the total picture looks like the following:

The value of μ_R is 7 for Gd while zero for Y. As we see from the above schematic figure, net magnetic moment would be dominated by rate earth ions when μ_R is greater than $5/3$.

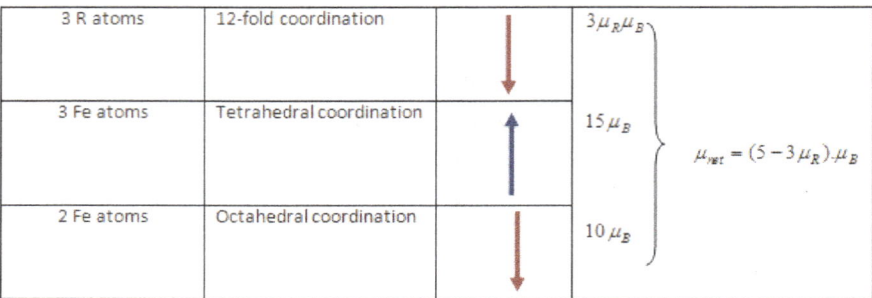

3 R atoms	12-fold coordination		$3\mu_R\mu_B$	
3 Fe atoms	Tetrahedral coordination		$15\mu_B$	$\mu_{net} = (5 - 3\mu_R).\mu_B$
2 Fe atoms	Octahedral coordination		$10\mu_B$	

This is dependent upon the temperature which governs the coupling between rare earth and Fe ions. Typically the net magnetic moment drops as the temperature increases, especially for strongly magnetic ions like Gd, Tb and Dy. Gd-doped garnet of composition $Y_{1.2}Gd_{1.8}Fe_5O_{12}$ has a rather stable saturation magnetization for a wide temperature range centered around ~50°C.

Garnets can be quite useful materials in microwave applications because of their high electrical resistivity and hence lower losses around microwave frequencies. The material is also easy to synthesize in either of bulk polycrystalline ceramic, single crystal or thin film forms. The structural parameters as well as magnetic properties can be tuned by tailoring the composition of the material.

Properties of Ferrite Ceramics

As discussed above, ferrites can have a broad spectrum of properties depending upon the type of ferrite and compositions. The following table shows some properties of common magnetic ceramics. These values are quite dependent upon the microstructure and processing history of the material. For comparison, values for some common magnetic metals and alloys are also given:

Material	Magnetic permeability, μ_R	Coercive Field (H_c, A.m⁻)	Remanence (B_s, T)	Curie temperature (T_c, °C)	Resistivity (p, Ω,m)
Soft materials					
Fe	150	80	2.16	770	10×10^{-8}
Fe-4%Si	2000	30	1.93	690	60×10^{-8}
Mn-Zn ferrites	500-10,000	5-100	0.35-0.50	90-280	0.01-1
Ni-Zn ferrites	10-2000	15-1600	0.10-0.40	90-500	10^3-10^7
Hard materials					
Medium Carbon steel		4.4	0.9	770	
Isotropic Barium Hexaferrite		200	0.3	450	
Oriented Barium Hexaferrite		320	0.4	450	

Soft Ferrites

Soft ferrites, as we explained earlier, are materials which are easy to magnetize or demagnetize i.e. materials with low coercive field strengths and thus so that they can reverse the direction in alternating fields without dissipating much energy since the area of B-H (or M-H) loop is small.

Typical soft ferrites used in transformer or electromagnetic cores contain nickel, zinc, and/or manganese based ferrites. These materials also have higher resistivities than typical ferromagnetic metals, of the order of 10^{-1} to $10^{6}\Omega.m$, which leads to low eddy currents in the core, another source of energy loss.

Because of their comparatively low losses at high frequencies, they are also extensively used in the cores of RF transformers and inductors in applications such as switched-mode power supplies (SMPS). The most common soft ferrites are $Mn_xZn_{(1-x)}Fe_2O_4$, $Ni_xZn_{(1-x)}Fe_2O_4$. Ferrites of Ni-Zn show higher resistivity than those containing Mn-Zn, and are, therefore, more suitable for frequencies above 1 MHz. Mn-Zn ferrites, in comparison, have higher permeability and saturation induction.

The properties of soft ferrites can be tailored by compositional modifications. For instance, in $Mn_{1-x}Zn_xFe_2O_4$ and $Ni_xZn_{(1-x)}Fe_2O_4$, increasing the Zn content leads to an increase in the magnetic permeability just before the magnetic transition but at the same the magnetic transition temperature also decreases. The increase in magnetic permeability near magnetic transition has been attributed to reduced magnetic anisotropy. The increase in relative permeability is about an order of magnitude in $Mn_{1-x}Zn_xFe_2O_4$ for doping levels up to 50 at % and about 2-3 orders of magnitude in $Ni_{1-x}Zn_xFe_2O_4$ for doping levels up to 70 at %.

Change in grain size also has a profound effect on the relative permeability with permeability increasing with increasing grain size. This is related to the decrease in the grain boundary concentration resulting in less pinning of domain walls by grain boundaries and hence facilitating easy magnetic switching. Electrical resistivity of ferrites is again composition dependent. Electrical conduction in ceramics takes place by hopping of electrons between say two valence states of an ion. In ferrites, d-group elements are susceptible to valence fluctuations. For example Mn-Zn ferrites are more susceptible to valence fluctuations of Mn and Fe as compared to Ni-Zn ferrites. This is also controlled very strongly by processing conditions such as firing temperatures, atmosphere and rate of cooling after sintering.

For example in $Ni_{1-x}Zn_xFe_{2+\alpha}O_{4-\beta}$, if all of the iron is present in 3+ valence state, then $\alpha=0$. Any increase in the iron content i.e. $\alpha > 0$ is compensated by the formation of Fe^{2+} which creates favourable conditions for electron hopping between Fe^{3+} and Fe^{2+} promoting n-type conduction. On the other hand, deficiency of iron i.e. $\alpha < 0$ is usually compensated by oxygen vacancies, resulting in a large increase in the resistivity, about 8 orders of magnitude, as shown below.

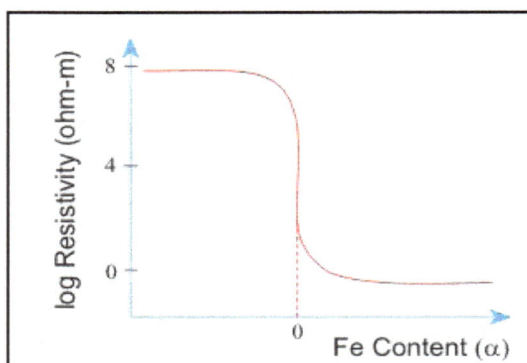

Resistivity variation in $Ni_{1-x}Zn_xFe_{2+\alpha}O_{4-\beta i}$ as a function of Fe content.

In addition, the resistivity of nickel ferrites is also increased by addition of small amounts of Cobalt. Reduction of cobalt from Co^{3+} to Co^{2+} state minimizes the reduction of iron to Fe^{2+} state. Since Co ions are sparsely located in the lattice, hopping of electrons between Co^{2+} and Co^{3+} states is minimal. Another method of increasing the resistivity in polycrystalline ferrites can be via grain boundary modification either by preferential oxidation of grain boundaries or by addition of additives like CaO or SiO_2 so that Fe and Ca ions are incorporated into ferrite regions closer to the grain boundaries. Both of these approaches make grain boundaries more resistive than the grains.

Another interest in soft ferrites is the frequency dependence if their properties such as permeability and loss. These properties are very useful for microwave applications, switch mode power supplies, inductors and other high frequency broadband applications.

Properties of Hard Ferrites

Hard ferrites or hard magnets exhibit high coercive fields, typically above 150 kA.m[1] and are often called permanent magnets. This is because these materials are able to withstand any demagnetizing effects that may arise either internally or externally.

In addition to magnetization or remanence (B_r) and coercive field, a permanent magnet is often characterized by product B.H *i.e.* area under the magnetic hysteresis curve.

Remanence in materials like hexaferrites which are strongly anisotropic can be affected by processing. Since the effect of magnetic anisotropy in a polycrystalline ceramic is small, the material can be synthesized under application of a magnetic field thereby aligning moments along c-axis in the grains giving larger remanence. Similarly, coercivity is a strong function of grain size and it is found for hexaferrites that it is maximum for grain sizes of about 1 μm.

Applications of Magnetic Ceramics

• In electronic inductors, transformers and electromagnets

   Soft ferrites like Mn-Zn and Ni-Zn ferrites are used as core materials in these applications in the frequencies ranging from a 100 kHz to 100 MHz. Typically these ferrites have high electrical resistance which results in very low eddy current losses. Most common radio magnets, including those used in loudspeakers, are ferrite magnets. Ferrite magnets have largely displaced Alnico magnets in these applications.

Ferrites are also used for power transformers which are used to transmit either over a single frequency or within a range such as in ultrasonic generators. For high frequency applications, upto about 5 MHz, Ni-Zn ferrites are useful while for frequencies upto 100 kHz, Mn-Zn ferrites are preferred due to their higher permeabilities.

- Equipment shielding

Here, due to their high impedance to high frequency currents, ferrite components of Ni-Zn and Mn-Zn ferrites are able to prevent high frequency electrical noise due to electromagnetic interference from exiting or entering the equipment.

- Data storage (e.g. magnetic recording tapes and hard disks)

In the magnetic tapes, elongated 0.2-.5 μm long hard magnetic oxide particles of γ - Fe_2O_4 are embedded in nonmagnetic binder. The particles have single domains magnetized along their major axes which are aligned in the plane of the tape. The coercive fields are typically between 50-100 kA.m^{-1}. In magnetic hard-disks, core element is produced by forming several layers of materials (nonmagnetic underlayer, magnetic layer, overcoat, plus layer of lubricants on a nonmagnetic disk substrate). Here, the read/write head is not in direct contact with the hard disk (in contrast to floppy disk) due to an air bearing ($\approx$ 50 nm); air flow is caused by the relative velocity between disk and head. These memories have high storage density of about 10 GB in^{-2} and short access time.

Early computer memories stored data in the residual magnetic fields of hard ferrite cores, which were assembled into arrays of core memory. Ferrite powders are used in the coatings of magnetic recording tapes. One such type of material is iron (III) oxide.

- Absorbing materials

In stealth aircrafts, ferrite particles are used as a component of radar-absorbing materials or coatings and in the absorption tile lining in the rooms used for electromagnetic compatibility measurements.

- Microwave applications in the frequency ranges of 1-300 GHz

Materials like Mg-ferrites, Li-doped Ferrites and garnets are used for such applications such as phase shifters, circulators and isolators.

References

- F. K. Lotgering, P. H. G. M. Vromans, M. A. H. Huyberts, "Permanent-magnet material obtained by sintering the hexagonal ferrite W=$BaFe_2Fe_{16}O_{27}$", Journal of Applied Physics, vol. 51, pp. 5913-5918, 1980

- A. Herczynski (2013). "Bound charges and currents" (PDF). American Journal of Physics. 81 (3): 202–205. Bibcode:2013AmJPh..81..202H. doi:10.1119/1.4773441

- Feynman, Richard P.; Leighton, Robert B.; Sands, Matthew (2006). The Feynman Lectures on Physics. 2. ISBN 0-8053-9045-6

- Cullity, B. D.; Graham, C. D. (2008). Introduction to Magnetic Materials (2nd ed.). Wiley-IEEE Press. p. 103. ISBN 0-471-47741-9

- Drakos, Nikos; Moore, Ross; Young, Peter (2002). "Landau diamagnetism". Electrons in a magnetic field. Retrieved 02, May 2020

- Liu, Yuanming; Zhu, Da-Ming; Strayer, Donald M.; Israelsson, Ulf E. (2010). "Magnetic levitation of large water droplets and mice". Advances in Space Research. 45 (1): 208–213. Bibcode:2010AdSpR..45..208L. doi:10.1016/j.asr.2009.08.033

- "Fun with diamagnetic levitation". ForceField. 02-12-2008. Archived from the original on February 12, 2008. Retrieved 15, June 2020

- Jackson, Roland (21 July 2014). "John Tyndall and the Early History of Diamagnetism". Annals of Science: 4. doi:10.1080/00033790.2014.929743. Retrieved 19, May 2020

- Buxton, Richard B. (2002). Introduction to functional magnetic resonance imaging. Cambridge University Press. p. 136. ISBN 0-521-58113-3

- "Search results matching 'magnetic moment'". CODATA internationally recommended values of the Fundamental Physical Constants. National Institute of Standards and Technology. Retrieved 11, May 2020

- Beatty, Bill (2005). "Neodymium supermagnets: Some demonstrations—Diamagnetic water". Science Hobbyist. Retrieved 01, May 2020

- Steiner, Marcus (2004). Micromagnetism and Electrical Resistance of Ferromagnetic Electrodes for Spin Injection Devices. Cuvillier Verlag. p. 6. ISBN 3-86537-176-0

- Ullah, Zaka; Atiq, Shahid; Naseem, Shahzad (2013). "Influence of Pb doping on structural, electrical and magnetic properties of Sr-hexaferrites". Journal of Alloys and Compounds. 555: 263–267. doi:10.1016/j.jallcom. Retrieved 09, April 2020

- Furlani, Edward P. (2001). Permanent Magnet and Electromechanical Devices: Materials, Analysis, and Applications. Academic Press. p. 140. ISBN 0-12-269951-3

Various Types of Ceramic Materials

There are different types of ceramic materials such as zirconium dioxide, silicon nitride, nanoceramic, coade stone, ceramic matrix composite and transparent ceramics. Zirconium dioxide is a white crystalline oxide of zirconium. All these diverse aspects of these types of ceramics have been carefully analyzed in this chapter.

Zirconium Dioxide

Zirconium dioxide (ZrO_2), sometimes known as zirconia, is a white crystalline oxide of zirconium. Its most naturally occurring form, with a monoclinic crystalline structure, is the mineral baddeleyite. A dopant stabilized cubic structured zirconia, cubic zirconia, is synthesized in various colours for use as a gemstone and a diamond simulant.

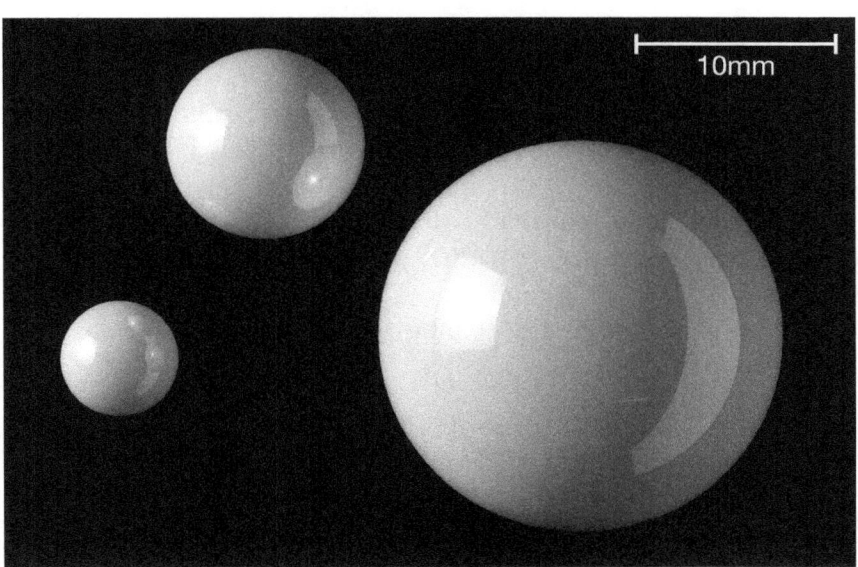

Bearing balls.

Production, Chemical Properties, Occurrence

Zirconia is produced by calcining zirconium compounds, exploiting its high thermal stability.

Structure

Three phases are known: monoclinic <1,170 °C, tetragonal 1,170–2,370 °C, and cubic >2,370 °C. The trend is for higher symmetry at higher temperatures, as is usually the case. A few percentage of the oxides of calcium or yttrium stabilize the cubic phase. The very rare mineral tazheranite

(Zr,Ti,Ca)O$_2$ is cubic. Unlike TiO$_2$, which features six-coordinate Ti in all phases, monoclinic zirconia consists of seven-coordinate zirconium centres. This difference is attributed to the larger size of Zr atom relative to the Ti atom.

Chemical Reactions

Zirconia is chemically unreactive. It is slowly attacked by concentrated hydrofluoric acid and sulfuric acid. When heated with carbon, it converts to zirconium carbide. When heated with carbon in the presence of chlorine, it converts to zirconium tetrachloride. This conversion is the basis for the purification of zirconium metal and is analogous to the Kroll process.

Engineering Properties

Zirconium dioxide is one of the most studied ceramic materials. ZrO$_2$ adopts a monoclinic crystal structure at room temperature and transitions to tetragonal and cubic at higher temperatures. The volume expansion caused by the cubic to tetragonal to monoclinic transformation induces large stresses, and these stresses cause ZrO$_2$ to crack upon cooling from high temperatures. When the zirconia is blended with some other oxides, the tetragonal and/or cubic phases are stabilized. Effective dopants include magnesium oxide (MgO), yttrium oxide (Y$_2$O$_3$, yttria), calcium oxide (CaO), and cerium(III) oxide (Ce$_2$O$_3$).

Zirconia is often more useful in its phase 'stabilized' state. Upon heating, zirconia undergoes disruptive phase changes. By adding small percentages of yttria, these phase changes are eliminated, and the resulting material has superior thermal, mechanical, and electrical properties. In some cases, the tetragonal phase can be metastable. If sufficient quantities of the metastable tetragonal phase is present, then an applied stress, magnified by the stress concentration at a crack tip, can cause the tetragonal phase to convert to monoclinic, with the associated volume expansion. This phase transformation can then put the crack into compression, retarding its growth, and enhancing the fracture toughness. This mechanism is known as transformation toughening, and significantly extends the reliability and lifetime of products made with stabilized zirconia.

The ZrO$_2$ band gap is dependent on the phase (cubic, tetragonal, monoclinic, or amorphous) and preparation methods, with typical estimates from 5–7 eV (0.80–1.12 aJ).

A special case of zirconia is that of tetragonal zirconia polycrystal, or TZP, which is indicative of polycrystalline zirconia composed of only the metastable tetragonal phase.

Uses

The main use of zirconia is in the production of ceramics, with other uses including as a protective coating on particles of titanium dioxide pigments, as a refractory material, in insulation, abrasives and enamels. Stabilized zirconia is used in oxygen sensors and fuel cell membranes because it has the ability to allow oxygen ions to move freely through the crystal structure at high temperatures. This high ionic conductivity (and a low electronic conductivity) makes it one of the most useful electroceramics. Zirconium dioxide is also used as the solid electrolyte in electrochromic devices.

Zirconia is a precursor to the electroceramic lead zirconate titanate (*PZT*), which is a high-K dielectric, which is found in myriad components.

Niche Uses

The very low thermal conductivity of cubic phase of zirconia also has led to its use as a thermal barrier coating, or TBC, in jet and diesel engines to allow operation at higher temperatures. Thermodynamically, the higher the operation temperature of an engine, the greater the possible efficiency. Another low thermal conductivity use is a ceramic fiber insulation for crystal growth furnaces, fuel cell stack insulation and infrared heating systems.

This material is also used in dentistry in the manufacture of 1) subframes for the construction of dental restorations such as crowns and bridges, which are then veneered with a conventional feldspathic porcelain for aesthetic reasons, or of 2) strong, extremely durable dental prostheses constructed entirely from monolithic zirconia, with limited but constantly improving aesthetics.

Zirconia is used to make ceramic knives. Because of its hardness, zirconia based cutlery stays sharp longer than a stainless steel equivalent.

Due to its infusibility and brilliant luminosity when incandescent, it was used as an ingredient of sticks for limelight.

Zirconia has been proposed to electrolyze carbon monoxide and oxygen from the atmosphere of Mars to provide both fuel and oxidizer that could be used as a store of chemical energy for use with surface transportation on Mars. Carbon monoxide/oxygen engines have been suggested for early surface transportation use as both carbon monoxide and oxygen can be straightforwardly produced by zirconia electrolysis without requiring use of any of the Martian water resources to obtain hydrogen, which would be needed for the production of methane or any hydrogen-based fuels.

Zirconia is also a potential high-k dielectric material with potential applications as an insulator in transistors.

Zirconia is also employed in the deposition of optical coatings; it is a high-index material usable from the near-UV to the mid-IR, due to its low absorption in this spectral region. In such applications, it is typically deposited by PVD.

An example of zirconium dioxide (ZrO_2) use in a consumer product is a line of the Omega Speedmaster Moonwatch collection, where the watch housings feature laser-engraved "ZrO_2" insignia.

Diamond Simulant

Single crystals of the cubic phase of zirconia are commonly used as diamond simulant in jewellery. Like diamond, cubic zirconia has a cubic crystal structure and a high index of refraction. Visually discerning a good quality cubic zirconia gem from a diamond is difficult, and most jewellers will have a thermal conductivity tester to identify cubic zirconia by its low thermal conductivity (diamond is a very good thermal conductor). This state of zirconia is commonly called *cubic zirconia*,

CZ, or *zircon* by jewellers, but the last name is not chemically accurate. Zircon is actually the mineral name for naturally occurring zirconium silicate ($ZrSiO_4$).

Brilliant-cut cubic zirconia.

Boron Carbide

Boron carbide (chemical formula approximately B_4C) is an extremely hard boron–carbon ceramic, and covalent material used in tank armor, bulletproof vests, engine sabotage powders, as well as numerous industrial applications. With a Vickers Hardness of >30 MPa, it is one of the hardest known materials, behind cubic boron nitride and diamond.

Boron carbide was discovered in 19th century as a by-product of reactions involving metal borides, however, its chemical formula was unknown. It was not until the 1930s that the chemical composition was estimated as B_4C. There remained, however, controversy as to whether or not the material had this exact 4:1 stoichiometry, as in practice the material is always slightly carbon-deficient with regard to this formula, and X-ray crystallography shows that its structure is highly complex, with a mixture of C-B-C chains and B_{12} icosahedra. These features argued against a very simple exact B_4C empirical formula. Because of the B_{12} structural unit, the chemical formula of "ideal" boron carbide is often written not as B_4C, but as $B_{12}C_3$, and the carbon deficiency of boron carbide described in terms of a combination of the $B_{12}C_3$ and $B_{12}CBC$ units.

The ability of boron carbide to absorb neutrons without forming long-lived radionuclides makes it attractive as an absorbent for neutron radiation arising in nuclear power plants and from anti-personnel neutron bombs. Nuclear applications of boron carbide include shielding, control rod and shut down pellets. Within control rods, boron carbide is often powdered, to increase its surface area.

Crystal Structure

Boron carbide has a complex crystal structure typical of icosahedron-based borides. There, B_{12} icosahedra form a rhombohedral lattice unit (space group: *R3m* (No. 166), lattice constants: $a = 0.56$ nm and $c = 1.212$ nm) surrounding a C-B-C chain that resides at the center of the unit cell, and both carbon atoms bridge the neighboring three icosahedra. This structure is layered: the B_{12} icosahedra and bridging carbons form a network plane that spreads parallel to the *c*-plane and stacks along the *c*-axis. The lattice has two basic structure units – the B_{12} icosahedron and the B_6 octahedron. Because of the small size of the B_6 octahedra, they cannot interconnect. Instead, they bond to the B_{12} icosahedra in the neighboring layer, and this decreases bonding strength in the *c*-plane.

Unit cell of B_4C. The green sphere and icosahedra consist of boron atoms, and black spheres are carbon atoms.

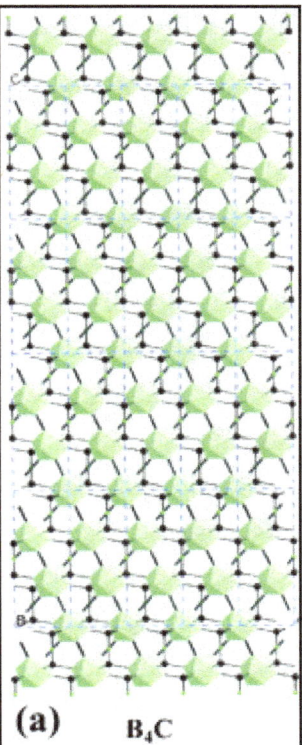

Fragment of the B_4C crystal structure.

Because of the B_{12} structural unit, the chemical formula of "ideal" boron carbide is often written not as B_4C, but as $B_{12}C_3$, and the carbon deficiency of boron carbide described in terms of a combination of the $B_{12}C_3$ and $B_{12}C_2$ units. Some studies indicate the possibility of incorporation of one or more carbon atoms into the boron icosahedra, giving rise to formulas such as $(B_{11}C)CBC = B_4C$ at the carbon-heavy end of the stoichiometry, but formulas such as $B_{12}(CBB) = B_{14}C$ at the boron-rich end. "Boron carbide" is thus not a single compound, but a family of compounds of different compositions. A common intermediate, which approximates a commonly found ratio of elements, is

$B_{12}(CBC) = B_{6.5}C$. Quantum mechanical calculations have demonstrated that configurational disorder between boron and carbon atoms on the different positions in the crystal determines several of the materials properties. In particular the crystal symmetry of the B_4C composition and the non-metallic electrical character of the $B_{13}C_2$ composition.

Properties

Boron carbide is known as a robust material having high hardness, high cross section for absorption of neutrons (i.e. good shielding properties against neutrons), stability to ionizing radiation and most chemicals. Its Vickers hardness (38 GPa), Elastic Modulus (460 GPa) and fracture toughness (3.5 MPa·m$^{1/2}$) approach the corresponding values for diamond (1150 GPa and 5.3 MPa·m$^{1/2}$).

As of 2015, boron carbide is the third hardest substance known, after diamond and cubic boron nitride, earning it the nickname "black diamond".

Semiconductor Properties

Boron carbide is a semiconductor, with electronic properties dominated by hopping-type transport. The energy band gap depends on composition as well as the degree of order. The band gap is estimated at 2.09 eV, with multiple mid-bandgap states which complicate the photoluminescence spectrum. The material is typically p-type.

Preparation

Boron carbide was first synthesized by Henri Moissan in 1899, by reduction of boron trioxide either with carbon or magnesium in presence of carbon in an electric arc furnace. In the case of carbon, the reaction occurs at temperatures above the melting point of B_4C and is accompanied by liberation of large amount of carbon monoxide:

$$2\,B_2O_3 + 7\,C \rightarrow B_4C + 6\,CO$$

If magnesium is used, the reaction can be carried out in a graphite furnace, and the magnesium byproducts are removed by treatment with acid.

Uses

- Padlocks.
- Personal and vehicle anti-ballistic armor plating.
- Grit blasting nozzles.
- High-pressure water jet cutter nozzles.
- Scratch and wear resistant coatings.
- Cutting tools and dies.
- Abrasives.

- Neutron absorber in nuclear reactors.

- Metal matrix composites.

- High energy fuel for solid fuel Ramjets.

- In brake linings of vehicles.

Silicon Nitride

Silicon nitride is a chemical compound of the elements silicon and nitrogen, with the formula Si_3N_4. It is a white, high-melting-point solid that is relatively chemically inert, being attacked by dilute HF and hot H_2SO_4. It is very hard (8.5 on the mohs scale). It is the most thermodynamically stable of the silicon nitrides. Hence, Si_3N_4 is the most commercially important of the silicon nitrides and is generally understood as what is being referred to where the term "silicon nitride" is used.

Production

The material is prepared by heating powdered silicon between 1300 °C and 1400 °C in an atmosphere of nitrogen:

$$3\,Si + 2\,N_2 \rightarrow Si_3N_4$$

The silicon sample weight increases progressively due to the chemical combination of silicon and nitrogen. Without an iron catalyst, the reaction is complete after several hours (~7), when no further weight increase due to nitrogen absorption (per gram of silicon) is detected. In addition to Si_3N_4, several other silicon nitride phases (with chemical formulas corresponding to varying degrees of nitridation/Si oxidation state) have been reported in the literature, for example, the gaseous disilicon mononitride (Si_2N); silicon mononitride (SiN), and silicon sesquinitride (Si_2N_3), each of which are stoichiometric phases. As with other refractories, the products obtained in these high-temperature syntheses depends on the reaction conditions (e.g. time, temperature, and starting materials including the reactants and container materials), as well as the mode of purification. However, the existence of the sesquinitride has since come into question.

It can also be prepared by diimide route:

$$SiCl_4 + 6\,NH_3 \rightarrow Si(NH)_2 + 4\,NH_4Cl(s) \quad \text{at 0 °C}$$

$$3Si(NH)_2 \rightarrow Si_3N_4 + N_2 + 3\,H_2(g) \quad \text{at 1000 °C}$$

Carbothermal reduction of silicon dioxide in nitrogen atmosphere at 1400–1450 °C has also been examined:

$$3\,SiO_2 + 6\,C + 2\,N_2 \rightarrow Si_3N_4 + 6\,CO$$

The nitridation of silicon powder was developed in the 1950s, following the "rediscovery" of silicon nitride and was the first large-scale method for powder production. However, use of low-purity raw silicon caused contamination of silicon nitride by silicates and iron. The diimide decomposition results in amorphous silicon nitride, which needs further annealing under nitrogen at 1400–1500 °C to convert it to crystalline powder; this is now the second-most important route for commercial production. The carbothermal reduction was the earliest used method for silicon nitride production and is now considered as the most-cost-effective industrial route to high-purity silicon nitride powder.

Electronic-grade silicon nitride films are formed using chemical vapor deposition (CVD), or one of its variants, such as plasma-enhanced chemical vapor deposition (PECVD):

$$3 \, SiH_4(g) + 4 \, NH_3(g) \rightarrow Si_3N_4(s) + 12 \, H_2(g)$$

$$3 \, SiCl_4(g) + 4 \, NH_3(g) \rightarrow Si_3N_4(s) + 12 \, HCl(g)$$

$$3 \, SiCl_2H_2(g) + 4 \, NH_3(g) \rightarrow Si_3N_4(s) + 6 \, HCl(g) + 6 \, H_2(g)$$

For deposition of silicon nitride layers on semiconductor (usually silicon) substrates, two methods are used:

1. Low pressure chemical vapor deposition (LPCVD) technology, which works at rather high temperature and is done either in a vertical or in a horizontal tube furnace, or

2. Plasma-enhanced chemical vapor deposition (PECVD) technology, which works at rather low temperature and vacuum conditions

The lattice constants of silicon nitride and silicon are different. Therefore, tension or stress can occur, depending on the deposition process. Especially when using PECVD technology this tension can be reduced by adjusting deposition parameters.

Silicon nitride nanowires can also be produced by sol-gel method using carbothermal reduction followed by nitridation of silica gel, which contains ultrafine carbon particles. The particles can be produced by decomposition of dextrose in the temperature range 1200–1350 °C. The possible synthesis reactions are:

$$SiO_2(s) + C(s) \rightarrow SiO(g) + CO(g) \quad and$$

$$3 \, SiO(g) + 2 \, N_2(g) + 3 \, CO(g) \rightarrow Si_3N_4(s) + 3 \, CO_2(g) \quad or$$

$$3 \, SiO(g) + 2 \, N_2(g) + 3 \, C(s) \rightarrow Si_3N_4(s) + 3 \, CO(g).$$

Processing

Silicon nitride is difficult to produce as a bulk material—it cannot be heated over 1850 °C, which is well below its melting point, due to dissociation to silicon and nitrogen. Therefore, application of conventional hot press sintering techniques is problematic. Bonding of silicon nitride powders can be achieved at lower temperatures through adding additional materials (sintering aids or "binders") which commonly induce a degree of liquid phase sintering. A cleaner alternative is to use spark plasma sintering where heating is conducted very rapidly (seconds) by passing pulses of electric current through the compacted powder. Dense silicon nitride compacts have been obtained by this techniques at temperatures 1500–1700 °C.

Crystal Structure and Properties

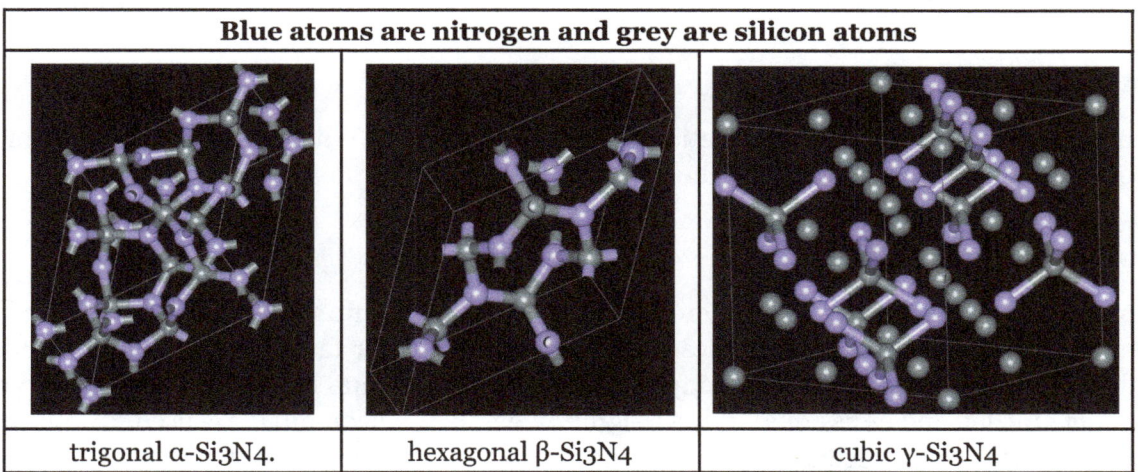

Blue atoms are nitrogen and grey are silicon atoms		
trigonal α-Si3N4.	hexagonal β-Si3N4	cubic γ-Si3N4

There exist three crystallographic structures of silicon nitride (Si3N4), designated as α, β and γ phases. The α and β phases are the most common forms of Si3N4, and can be produced under normal pressure condition. The γ phase can only be synthesized under high pressures and temperatures and has a hardness of 35 GPa.

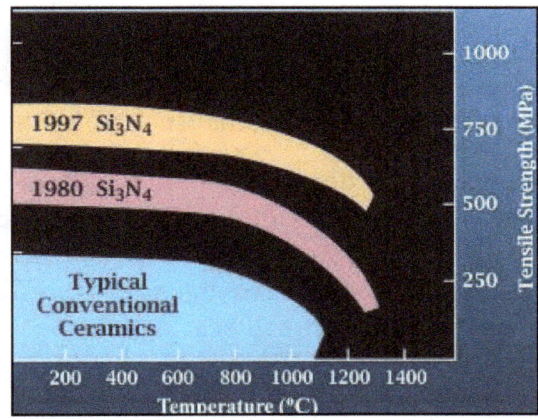

The α- and β-Si3N4 have trigonal (Pearson symbol hP28, space group P31c, No. 159) and hexagonal (hP14, P6$_3$, No. 173) structures, respectively, which are built up by corner-sharing SiN4 tetrahedra. They can be regarded as consisting of layers of silicon and nitrogen atoms in the sequence ABAB... or ABCDABCD... in β-Si3N4 and α-Si3N4, respectively. The AB layer is the same in the α and β phases, and the CD layer in the α phase is related to AB by a c-glide plane. The Si3N4 tetrahedra in β-Si3N4 are interconnected in such a way that tunnels are formed, running parallel with the c axis of the unit cell. Due to the c-glide plane that relates AB to CD, the α structure contains cavities instead of tunnels. The cubic γ-Si3N4 is often designated as c modification in the literature, in analogy with the cubic modification of boron nitride (c-BN). It has a spinel-type structure in which two silicon atoms each coordinate six nitrogen atoms octahedrally, and one silicon atom coordinates four nitrogen atoms tetrahedrally.

The longer stacking sequence results in the α-phase having higher hardness than the β-phase. However, the α-phase is chemically unstable compared with the β-phase. At high temperatures when a liquid phase is present, the α-phase always transforms into the β-phase. Therefore, β-Si3N4 is the major form used in Si3N4 ceramics.

Applications

In general, the main issue with applications of silicon nitride has not been technical performance, but cost. As the cost has come down, the number of production applications is accelerating.

Automobile Industry

One of the major applications of sintered silicon nitride is in automobile industry as a material for engine parts. Those include, in diesel engines, glowplugs for faster start-up; precombustion chambers (swirl chambers) for lower emissions, faster start-up and lower noise; turbocharger for reduced engine lag and emissions. In spark-ignition engines, silicon nitride is used for rocker arm pads for lower wear, turbocharger for lower inertia and less engine lag, and in exhaust gas control valves for increased acceleration. As examples of production levels, there is an estimated more than 300,000 sintered silicon nitride turbochargers made annually.

Bearings

Silicon nitride bearings are both full ceramic bearings and ceramic hybrid bearings with balls in ceramics and races in steel. Silicon nitride ceramics have good shock resistance compared to other ceramics. Therefore, ball bearings made of silicon nitride ceramic are used in performance bearings. A representative example is use of silicon nitride bearings in the main engines of the NASA's Space Shuttle.

Since silicon nitride ball bearings are harder than metal, this reduces contact with the bearing track. This results in 80% less friction, 3 to 10 times longer lifetime, 80% higher speed, 60% less weight, the ability to operate with lubrication starvation, higher corrosion resistance and higher operation temperature, as compared to traditional metal bearings. Silicon nitride balls weigh 79% less than tungsten carbide balls. Silicon nitride ball bearings can be found in high end automotive bearings, industrial bearings, wind turbines, motorsports, bicycles, rollerblades and skateboards. Silicon nitride bearings are especially useful in applications where corrosion, electric or magnetic fields prohibit the use of metals. For example, in tidal flow meters, where seawater attack is a problem, or in electric field seekers.

Si_3N_4 was first demonstrated as a superior bearing in 1972 but did not reach production until nearly 1990 because of challenges associated with reducing the cost. Since 1990, the cost has been reduced substantially as production volume has increased. Although Si3N4 bearings are still 2–5 times more expensive than the best steel bearings, their superior performance and life are justifying rapid adoption. Around 15–20 million Si3N4 bearing balls were produced in the U.S. in 1996 for machine tools and many other applications. Growth is estimated at 40% per year, but could be even higher if ceramic bearings are selected for consumer applications such as in-line skates and computer disk drives.

High-temperature Material

Silicon nitride has long been used in high-temperature applications. In particular, it was identified as one of the few monolithic ceramic materials capable of surviving the severe thermal shock and thermal gradients generated in hydrogen/oxygen rocket engines. To demonstrate this capability in a complex configuration, NASA scientists used advanced rapid prototyping technology to fabricate a one-inch-diameter, single-piece combustion chamber/nozzle (thruster) component. The thruster was hot-fire tested with hydrogen/oxygen propellant and survived five cycles including a 5-minute cycle to a 1320 °C material temperature.

In 2010 silicon nitride was used as the main material in the thrusters of the JAXA space probe Akatsuki.

Medical

Silicon nitride has many orthopedic applications. The material is also an alternative to PEEK (polyether ether ketone) and titanium, which are used for spinal fusion devices. It is silicon nitride's hydrophilic, microtextured surface that contributes to the materials strength, durability and reliability compared to PEEK and titanium.

Metal Working and Cutting Tools

The first major application of Si3N4 was abrasive and cutting tools. Bulk, monolithic silicon nitride is used as a material for cutting tools, due to its hardness, thermal stability, and resistance to wear. It is especially recommended for high speed machining of cast iron. Hot hardness, fracture toughness and thermal shock resistance mean that sintered silicon nitride can cut cast iron, hard steel and nickel based alloys with surface speeds up to 25 times quicker than those obtained with conventional materials such as tungsten carbide. The use of Si 3N4 cutting tools has had a dramatic effect on manufacturing output. For example, face milling of gray cast iron with silicon nitride inserts doubled the cutting speed, increased tool life from one part to six parts per edge, and reduced the average cost of inserts by 50%, as compared to traditional tungsten carbide tools.

Electronics

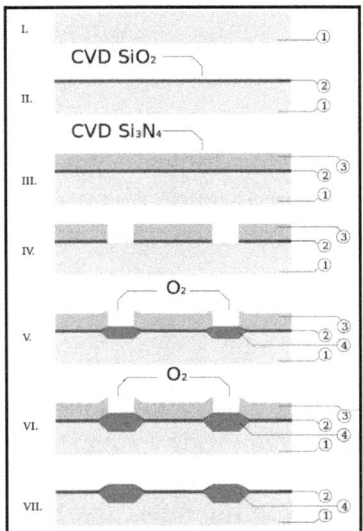

Example of local silicon oxidation through a Si_3N_4 mask.

Silicon nitride is often used as an insulator and chemical barrier in manufacturing integrated circuits, to electrically isolate different structures or as an etch mask in bulk micromachining. As a passivation layer for microchips, it is superior to silicon dioxide, as it is a significantly better diffusion barrier against water molecules and sodium ions, two major sources of corrosion and instability in microelectronics. It is also used as a dielectric between polysilicon layers in capacitors in analog chips.

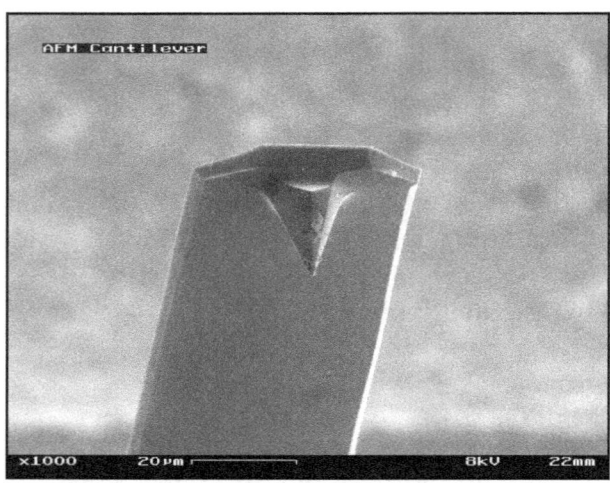
Si_3N_4 cantilever used in atomic force microscopes.

Silicon nitride deposited by LPCVD contains up to 8% hydrogen. It also experiences strong tensile stress, which may crack films thicker than 200 nm. However, it has higher resistivity and dielectric strength than most insulators commonly available in microfabrication (10^{16} Ω·cm and 10 MV/cm, respectively).

Not only silicon nitride, but also various ternary compounds of silicon, nitrogen and hydrogen (SiN_xH_y) are used insulating layers. They are plasma deposited using the following reactions:

$$2\ SiH4(g) + N2(g) \rightarrow 2\ SiNH(s) + 3\ H2(g)$$

$$SiH4(g) + NH3(g) \rightarrow SiNH(s) + 3\ H2(g)$$

These SiNH films have much less tensile stress, but worse electrical properties (resistivity 10^6 to 10^{15} Ω·cm, and dielectric strength 1 to 5 MV/cm).

Silicon nitride is also used in xerographic process as one of the layer of the photo drum. Silicon nitride is also used as an ignition source for domestic gas appliances. Because of its good elastic properties, silicon nitride, along with silicon and silicon oxide, is the most popular material for cantilevers — the sensing elements of atomic force microscopes.

History

The first reported preparation was in 1857 by Henri Etienne Sainte-Claire Deville and Friedrich Wöhler. In their method, silicon was heated in a crucible placed inside another crucible packed with carbon to reduce permeation of oxygen to the inner crucible. They reported a product they termed silicon nitride but without specifying its chemical composition. Paul Schuetzenberger first reported a product with the composition of the tetranitride, Si3N4, in 1879 that was obtained by heating silicon with brasque (a paste made by mixing charcoal, coal, or coke with clay which is then used to line crucibles) in a blast furnace. In 1910, Ludwig Weiss and Theodor Engelhardt heated silicon under pure nitrogen to produce Si3N4. E. Friederich and L. Sittig made Si_3N_4 in 1925 via carbothermal reduction under nitrogen, that is, by heating silica, carbon, and nitrogen at 1250–1300 °C.

Silicon nitride remained merely a chemical curiosity for decades before it was used in commercial applications. From 1948 to 1952, the Carborundum Company, Niagara Falls, New York, applied for several patents on the manufacture and application of silicon nitride. By 1958 Haynes (Union Carbide) silicon nitride was in commercial production for thermocouple tubes, rocket nozzles, and boats and crucibles for melting metals. British work on silicon nitride, started in 1953, was aimed at high-temperature parts of gas turbines and resulted in the development of reaction-bonded silicon nitride and hot-pressed silicon nitride. In 1971, the Advanced Research Project Agency of the US Department of Defense placed a US$17 million contract with Ford and Westinghouse for two ceramic gas turbines.

Even though the properties of silicon nitride were well known, its natural occurrence was discovered only in the 1990s, as tiny inclusions (about 2 μm × 0.5 μm in size) in meteorites. The mineral was named nierite after a pioneer of mass spectrometry, Alfred O. C. Nier. This mineral might have been detected earlier, again exclusively in meteorites, by Soviet geologists.

Nanoceramic

Nanoceramic is a type of nanoparticle that is composed of ceramics, which are generally classified as inorganic, heat-resistant, nonmetallic solids made of both metallic and nonmetallic compounds. The material offers unique properties. Macroscale ceramics are brittle and rigid and break upon impact. However, nanoceramics take on a larger variety of functions, including dielectric, ferroelectric, piezoelectric, pyroelectric, ferromagnetic, magnetoresistive, superconductive and electro-optical.

Nanoceramics were discovered in the early 1980s. They were formed using a process called sol-gel which mixes nanoparticles within a solution and gel to form the nanoparticle. Later methods involved sintering (pressure and heat). The material is so small that it has basically no flaws. Larger scale materials have flaws that render them brittle.

In 2014 researchers announced a lasering process involving polymers and ceramic particles to form a nanotruss. This structure was able to recover its original form after repeated crushing.

Properties

Nanoceramics have unique properties because of their size and molecular structure. These properties are often shown in terms of various electrical and magnetic physics phenomenons which include:

- Dielectric - An electrical insulator that can be polarized (having electrons aligned so that there is a negative and positive side of the compound) by an electric field to shorten the distance of electron transfer in an electric current.

- Ferroelectric - Dielectric materials that polarize in more than one direction (the negative and positive sides can be flipped via an electric field).

- Piezoelectric - Materials that accumulate an electrical charge under mechanical stress.

- Pyroelectric - Material that can produce a temporary voltage given a temperature change.

- Ferromagnetic - Materials that can to sustain a magnetic field after magnetization.

- Magnetoresistive - Materials that change electrical resistance under an external magnetic field.

- Superconductive - Materials that exhibit zero electric resistance when cooled to a critical temperature.

- Electro-optical - Materials that change optical properties under an electric field.

Nanotruss

Nanoceramic is more than 85% air and is very light, strong, flexible and durable. The fractal nanotruss is a nanostructure architecture made of alumina, or aluminum oxide. Its maximum compression is about 1 micron from a thickness of 50 nanometers. After its compression, it can revert to its original shape without any structural damage.

Synthesis

Sol-gel

One process for making nanoceramics varies is the sol-gel process, also known as chemical solution deposition. This involves a chemical solution, or the sol, made of nanoparticles in liquid phase and a precursor, usually a gel or polymer, made of molecules immersed in a solvent. The sol and gel are mixed to produce an oxide material which are generally a type of ceramic. The excess products (a liquid solvent) are evaporated. The particles desires are then heated in a process called densification to produce a solid product. This method could also be applied to produce a nanocomposite by heating the gel on a thin film to form a nanoceramic layer on top of the film.

Two-photon Lithography

This process uses a laser technique called two-photon lithography to etch out a polymer into a three-dimensional structure. The laser hardens the spots that it touches and leaves the rest unhardened. The unhardened material is then dissolved to produce a "shell". The shell is then coated with ceramic, metals, metallic glass, etc. In the finished state, the nanotruss of ceramic can be flattened and revert to its original state.

Sintering

In another approach sintering was used to consolidate nanoceramic powders using high temperatures. This resulted in a rough material that damages the properties of ceramics and requires more time to obtain an end product. This technique also limits the possible final geometries. Microwave sintering was developed to overcome such problems. Radiation is produced from a magnetron, which produces electromagnetic waves to vibrate and heat the powder. This method allows for heat to be instantly transferred across the entire volume of material instead of from the outside in.

The nanopowder is placed in an insulation box composed of low insulation boards to allow the microwaves to pass through it. The box increases temperature to aid absorption. Inside the boxes are suspectors that absorb microwaves at room temperature to initialize the sintering process. The microwave heats the suspectors to about 600 °C, sufficient to trigger the nanoceramics to absorb the microwaves.

History

In the early 1980s, the first nanoparticles, specifically nanoceramics were formed, using sol-gel. This process was replaced by sintering in the early 2000s and then by microwave sintering. None of these techniques proved suitable for large scale production.

In 2002, researchers tried to reverse engineer the microstructure of seashells to strengthen ceramics. They discovered that seashells' durability come from their "microarchitecture". Research began to focus on how ceramics could employ such an architecture.

In 2012 researchers replicated the sea sponge's structure using ceramics and the nanoarchitecture called nanotruss. As of 2015 the largest result is a 1mm cube. The lattice structure compresses up to 85% of its original thickness and can recover to its original form. These lattices are stabilized into triangles with cross-members for structural integrity and flexibility.

Applications

Medical technology used nanoceramics for bone repair. It has been suggested for areas including energy supply and storage, communication, transportation systems, construction and medical technology. Their electrical properties may allow energy to be transferred efficiencies approaching 100%. Nanotrusses may be eventually applicable for building materials, replacing concrete or steel.

Coade Stone

Father Thames, a Coade stone sculpture by John Bacon, in the grounds of Ham House.

Coade stone or *Lithodipyra* was stoneware that was often described as an artificial stone in the late 18th and early 19th centuries. It was used for moulding Neoclassical statues, architectural decorations and garden ornaments that were both of the highest quality and remain virtually weatherproof today. Produced by appointment to George III and the Prince Regent, it features on St George's Chapel, Windsor; The Royal Pavilion, Brighton; Carlton House, London; the Royal Naval College, Greenwich; and a large quantity was used in the refurbishment of Buckingham Palace in the 1820s.

Lithodipyra was first created around 1770 by Eleanor Coade who ran *Coade's Artificial Stone Manufactory*, *Coade and Sealy*, and *Coade* in Lambeth, London, from 1769 until her death in 1821, after which *Lithodipyra* continued to be manufactured by her last business partner William Croggon until 1833.

The recipe and techniques for producing Coade stone have been rediscovered by the team at Coade ltd., which now reproduce a range of Coade sculpture at their workshops in Wilton.

History

Lion Gate, an entrance into Kew Gardens, with its Coade stone lion statue on top.

In 1769 Mrs Coade bought Daniel Pincot's struggling artificial stone business at Kings Arms Stairs, Narrow Wall, Lambeth, a site now under the Royal Festival Hall. This business developed into *Coade's Artificial Stone Manufactory* with Eleanor in charge, such that within two years (1771) she fired Pincot for 'representing himself as the chief proprietor'.

Mrs Coade did not invent 'artificial stone' - various inferior quality precursors having been both patented and manufactured over the previous forty (or sixty) years - but she was probably responsible for perfecting both the clay recipe and the firing process. It is possible that Pincot's business was a continuation of that run nearby by Richard Holt, who had taken out two patents in 1722 for a kind of liquid metal or stone and another for making china without the use of clay, but there were many start-up 'artificial stone' businesses in the early 18th century of which only Mrs Coade's succeeded.

The company did well, and boasted an illustrious list of customers such as George III and members of the English nobility. In 1799 Mrs Coade appointed her cousin John Sealy (her mother's sister Mary's son), already working as a modeller, as a partner in her business, which then traded as 'Coade and Sealy' until his death in 1813 when it reverted to just 'Coade'.

In 1799 she opened a show room *Coade's Gallery* on *Pedlar's Acre* at the Surrey end of Westminster Bridge Road to display her products.

In 1813 Mrs Coade took on William Croggan from Grampound in Cornwall, a sculptor and distant relative by marriage (second cousin once removed). He managed the factory until her death eight years later in 1821 whereby he bought the factory from the executors for c. £4000. Croggan supplied a lot of Coade stone for Buckingham Palace; however, he went bankrupt in 1833 and died two years later. Trade declined, and production came to an end in the early 1840s.

In 2000 Coade ltd started reproducing Coade stone statues also creating new sculptures and architectural ornament, using the original recipes and methods of the eighteenth century.

The Material

Home of Eleanor Coade, Belmont House, in Lyme Regis, Dorset, with Coade stone ornamental façade.

Its colours varied from light grey to light yellow (or even beige) and its surface is best described as having a matte finish.

The ease with which the product could be moulded into complex shapes made it ideal for large statues, sculptures and sculptural façades. Moulds were often kept for many years, for repeated use. One-offs were clearly much more expensive to produce, as they had to carry the entire cost of creating the mould.

One of the more striking features of Coade stone is its incredible resistance to weathering, often faring better than most types of stone in London's harsh environment. Examples of Coade stone-work have survived very well; prominent examples are listed below, having survived without apparent wear and tear for 150 years.

As a material, Coade stone was replaced by Portland cement as a form of artificial stone and it appears to have been largely phased out by the 1840s.

Quality Controversy

Although Coade stone's reputation for both weather resistance and manufacturing quality is virtually untarnished, three sources describe Rossi's statue of George IV erected in the Royal Crescent, Brighton as "unable to withstand the weathering effects of sea-spray and strong wind: such that, by 1807 the fingers on the sculpture's left hand had been destroyed, and soon afterwards the whole right arm dropped off." By contrast however *Fashionable Brighton, 1820-1860* by Antony Dale (online) describes similar damage as 'wore badly' but does not attribute 'broken fingers, nose, mantle and arm on an unloved statue' to weathering or poor quality Coade stone. In 1819, after considerable complaints, the relic was removed and its present state is undocumented. A few works produced by Coade, mainly dating from the later period, have shown poor resistance to weathering due to a bad firing in the kiln where the material was not brought up to a sufficient temperature.

The Formula

Contrary to popular belief, the recipe for Coade stone still exists, and can be produced by Coade ltd. Rather than being based on cement (as concrete articles are), it is a ceramic material.

Its manufacture required special skills: extremely careful control and skill in kiln firing, over a period of days. This skill is even more remarkable when the potential variability of kiln temperatures at that time is considered. Mrs Coade's factory was the only really successful manufacturer.

The formula used was:

- 10% of grog
- 5-10% of crushed flint
- 5-10% fine quartz
- 10% crushed soda lime glass
- 60-70% Ball clay from Dorset and Devon

This mixture was also referred to as "fortified clay" which was then inserted after kneading into a kiln which would fire the material at a temperature of 1,100 °C for over four days.

A number of different variations of the recipe were used, depending on the size and fineness of detail in the work a different size and proportion of grog was used. In many pieces a combination of fine grogged Coade clay was used on the surface for detail, backed up by a more heavily grogged mixture for strength.

Examples

Over 650 pieces are still in existence worldwide.

The Red Lion, aka the South Bank Lion, on Westminster Bridge. Modelled by William F. Woodington and Grade II* listed by English Heritage.

Schomberg House circa 1850.

Captain William Bligh's Tomb surmounted by a breadfruit in a bowl.

London Lodge (1793), Highclere Castle, Hampshire. Brick but Coade stone dressed, and wings (1840), Highclere Castle, Hampshire, May 2014.

- Mrs Coade's country home, Belmont House in Lyme Regis, Dorset, displays examples of Coade stone on its façade

- The South Bank Lion at the south end of Westminster Bridge in central London originally stood atop the old Red Lion Brewery, on the Lambeth bank of the River Thames. When the brewery was demolished in 1950, to make way for the South Bank Site of the 1951 Festival of Britain, the Lion was taken down and moved to Station Approach Waterloo, painted red as the symbol of British Rail on high plinth. When removed, the initials of the sculptor William F. Woodington and the date, 24 May 1837, were discovered under one of its paws. The fine details still remain clear after 170 years of London's corrosive atmosphere, caused by heavy use of coal throughout the 19th and first half of the 20th centuries. The red paint was removed to reveal the fine Coade stone surface to view. In 1966, the statue was moved from outside Waterloo station to its current location

- Duff House Mausoleum, Wrack Woods, Banff, Aberdeenshire, Scotland. The second Earl of Fife built this mausoleum for his family tombs in 1791, possibly on the site of a Carmelite friary. Built before the Gothic Revival, this is an example of "Gothick" architecture. Typically of the Georgians the carvings, including the monument to the first Earl, are in ceramic Coade stone

- Nelson's Memorial at Burnham Thorpe

- St Botolph-without-Bishopsgate Church Hall, London, pair of statues of schoolchildren on the front of this former School House, replicas outside, listed originals now inside the Hall

- The statue and ornaments on Nelson's Column, Montreal, built 1809

- Britannia Monument in Great Yarmouth

- The Lion Gate, Mote Park, Roscommon, Ireland, built 1787

- Nelson's Pediment on the Old Royal Naval College, Greenwich, regarded by the Coade workers as the finest of all their work

- Twinings' first ever (and still operating) shop's frontispiece, in the Strand, London opposite the Royal Courts of Justice, rediscovered under soot after a century

- Schomberg House on Pall Mall, London

- Captain Bligh's tomb (in the churchyard of St Mary's Lambeth)

- Lord Hill's Column, Shrewsbury, Shropshire

- Rio de Janeiro's zoo entrance

- St Mary's Church gate, Tremadog, Gwynedd, Wales

- Richmond upon Thames. Two examples of the River God, one outside Ham House, the other in Terrace Gardens

- Buckingham Palace (in a section not open to the public)

- Castle Howard

- A couple of large ornate urns in the Italian Garden at Chiswick House, London

- Royal Pavilion in Brighton

- Imperial War Museum (sculptural reliefs above the entrance)

- The Buttermarket in Chichester, which was designed by John Nash (coat of arms engraved with "Coade & Sealey 1808")

- Saxham Hall, Suffolk has an Umbrello (shelter) constructed of Coade stone in the grounds

- The lion and unicorn statues over their respective gates into Kew Gardens

- Burton Constable Hall in Holderness, East Riding of Yorkshire displays 3 figures and a number of 'medallions' above the doors and windows of the Orangerie

- The Royal Arms of Queen Charlotte above 8 Argyle Street, Bath

- The keystone, featuring a carving of the head of Silenus, above the entry to The Rossborough Inn, a historic building at the University of Maryland, College Park, in the United States

- Edinburgh, three Coade Stone columns on Portobello Beach

Ceramic Matrix Composite

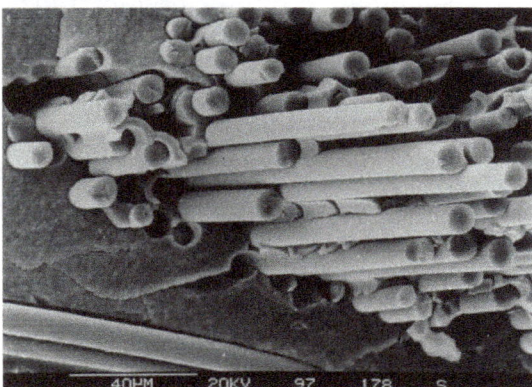

Fracture surface of a fiber-reinforced ceramic composed of SiC fibers and SiC matrix. The fiber pull-out mechanism shown is the key to CMC properties.

CMC shaft sleeves with outer diameters between 100 and 300 mm for ceramic slide bearings of pumps.

Ceramic matrix composites (CMCs) are a subgroup of composite materials as well as a subgroup of technical ceramics. They consist of ceramic fibres embedded in a ceramic matrix, thus forming a ceramic fibre reinforced ceramic (CFRC) material. The matrix and fibres can consist of any ceramic material, whereby carbon and carbon fibres can also be considered a ceramic material.

Introduction

The motivation to develop CMCs was to overcome the problems associated with the conventional technical ceramics like alumina, silicon carbide, aluminium nitride, silicon nitride or zirconia – they fracture easily under mechanical or thermo-mechanical loads because of cracks initiated by small defects or scratches. The crack resistance is – like in glass – very low. To increase the crack resistance or fracture toughness, particles (so-called monocrystalline *whiskers* or *platelets*) were embedded into the matrix. However, the improvement was limited, and the products have found application only in some ceramic cutting tools. So far only the integration of long multi-strand fibres has drastically increased the crack resistance, elongation and thermal shock resistance, and resulted in several new applications.

Carbon (C), special silicon carbide (SiC), alumina (Al_2O_3) and mullite (Al_2O_3–SiO_2) fibres are most commonly used for CMCs. The matrix materials are usually the same, that is C, SiC, alumina and mullite.

Generally, CMC names include a combination of *type of fibre/type of matrix*. For example, *C/C* stands for carbon-fibre-reinforced carbon (carbon/carbon), or *C/SiC* for carbon-fibre-reinforced silicon carbide. Sometimes the manufacturing process is included, and a C/SiC composite manufactured with the liquid polymer infiltration (LPI) process is abbreviated as *LPI-C/SiC*.

The important commercially available CMCs are C/C, C/SiC, SiC/SiC and Al_2O_3/Al_2O_3. They differ from conventional ceramics in the following properties, presented in more detail below:

- Elongation to rupture up to 1%

- Strongly increased fracture toughness

- Extreme thermal shock resistance

- Improved dynamical load capability

- Anisotropic properties following the orientation of fibers

Manufacture

The manufacturing processes usually consist of the following three steps:

1. Lay-up and fixation of the fibres, shaped as the desired component

2. Infiltration of the matrix material

3. Final machining and, if required, further treatments like coating or impregnation of the intrinsic porosity

The first and the last step are almost the same for all CMCs: In step one, the fibres, often named rovings, are arranged and fixed using techniques used in fibre-reinforced plastic materials, such as lay-up of fabrics, filament winding, braiding and knotting. The result of this procedure is called *fibre-preform* or simply *preform*.

For the second step, five different procedures are used to fill the ceramic matrix in between the fibres of the preform:

1. Deposition out of a gas mixture

2. Pyrolysis of a pre-ceramic polymer

3. Chemical reaction of elements

4. Sintering at a relatively low temperature in the range 1000–1200 °C

5. Electrophoretic deposition of a ceramic powder

Procedures one, two and three find applications with non-oxide CMCs, whereas the fourth one is used for oxide CMCs; combinations of these procedures are also practised. The fifth procedure is not yet established in industrial processes. All procedures have sub-variations, which differ in technical details. All procedures yield a porous material.

The third and final step of machining – grinding, drilling, lapping or milling – has to be done with diamond tools. CMCs can also be processed with a water jet, laser, or ultrasonic machining.

Ceramic Fibres

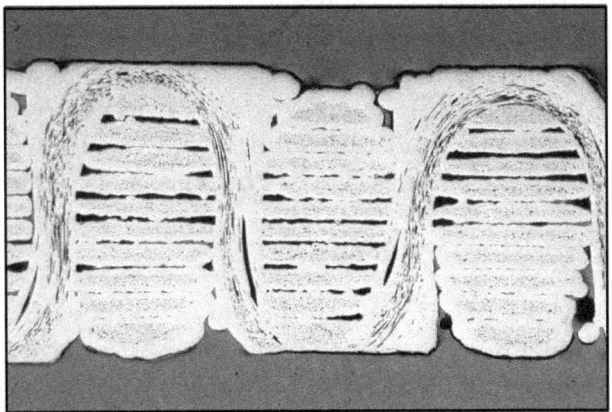

Micrograph of a SiC/SiC ceramic composite with a woven three-dimensional fibre structure.

Ceramic fibres in CMCs can have a polycrystalline structure, as in conventional ceramics. They can also be amorphous or have inhomogeneous chemical composition, which develops upon pyrolysis of organic precursors. The high process temperatures required for making CMCs preclude the use of organic, metallic or glass fibres. Only fibres stable at temperatures above 1000 °C can be used, such as fibres of alumina, mullite, SiC, zirconia or carbon. Amorphous SiC fibres have an elongation capability above 2% – much larger than in conventional ceramic materials (0.05 to 0.10%). The reason for this property of SiC fibres is that most of them contain additional elements like oxygen, titanium and/or aluminium yielding a tensile strength above 3 GPa. These enhanced elastic properties are required for various three-dimensional fibre arrangements in textile fabrication, where a small bending radius is essential.

Manufacturing Procedures

Matrix Deposition from a Gas Phase

Chemical vapour deposition (CVD) is well suited for this purpose. In the presence of a fibre pre-form, CVD takes place in between the fibres and their individual filaments and therefore is called chemical vapour infiltration (CVI). One example is the manufacture of C/C composites: a C-fibre preform is exposed to a mixture of argon and a hydrocarbon gas (methane, propane, etc.) at a pressure of around or below 100 kPa and a temperature above 1000 °C. The gas decomposes depositing carbon on and between the fibers. Another example is the deposition of silicon carbide, which is usually conducted from a mixture of hydrogen and methyl-trichlorosilane (MTS, CH_3SiCl_3; it is also common in silicone production). Under defined condition this gas mixture deposits fine and crystalline silicon carbide on the hot surface within the preform. This CVI procedure leaves a body with a porosity of about 10–15%, as access of reactants to the interior of the preform is increasingly blocked by deposition on the exterior.

Matrix Forming via Pyrolysis of C- and Si-containing Polymers

Hydrocarbon polymers shrink during pyrolysis, and upon outgassing form carbon with an amorphous, glass-like structure, which by additional heat treatment can be changed to a more graphite-like structure. Other special polymers, where some carbon atoms are replaced by silicon atoms, the so-called polycarbosilanes, yield amorphous silicon carbide of more or less stoichiometric composition. A large variety of such SiC-, SiNC-, or SiBNC-producing precursors already exist and more are being developed. To manufacture a CMC material, the fibre preform is infiltrated with the chosen polymer. Subsequent curing and pyrolysis yield a highly porous matrix, which is undesirable for most applications. Further cycles of polymer infiltration and pyrolysis are performed until the final and desired quality is achieved. Usually five to eight cycles are necessary. The process is called *liquid polymer infiltration* (LPI), or *polymer infiltration and pyrolysis* (PIP). Here also a porosity of about 15% is common due to the shrinking of the polymer. The porosity is reduced after every cycle.

Matrix Forming via Chemical Reaction

With this method, one material located between the fibres reacts with a second material to form the ceramic matrix. Some conventional ceramics are also manufactured by chemical reactions. For example, reaction-bonded silicon nitride (RBSN) is produced through the reaction of silicon powder with nitrogen, and porous carbon reacts with silicon to form reaction bonded silicon carbide, a silicon carbide which contains inclusions of a silicon phase. An example of CMC manufacture, which was introduced for the production of ceramic brake discs, is the reaction of silicon with a porous preform of C/C. The process temperature is above 1414 °C, that is above the melting point of silicon, and the process conditions are controlled such that the carbon fibres of the C/C-preform almost completely retain their mechanical properties. This process is called *liquid silicon infiltration* (LSI). Sometimes, and because of its starting point with C/C, the material is abbreviated as *C/C-SiC*. The material produced in this process has a very low porosity of about 3%.

Matrix Forming via Sintering

This process is used to manufacture oxide fibre/oxide matrix CMC materials. Since most ceramic fibres can not withstand the normal sintering temperatures of above 1600 °C, special precursor liquids are used to infiltrate the preform of oxide fibres. These precursors allow sintering, that is ceramic-forming processes, at temperatures of 1000–1200 °C. They are, for example, based on mixtures of alumina powder with the liquids tetra-ethyl-orthosilicate (as Si donor) and alumini-um-butylate (as Al donor), which yield a mullite matrix. Other techniques, such as sol-gel chemistry, are also used. CMCs obtained with this process usually have a high porosity of about 20%.

Matrix formed via electrophoresis

In the electrophoretic process, electrically charged particles dispersed in a special liquid are transported through an electric field into the preform, which has the opposite electrical charge polarity. This process is under development, and is not yet used industrially. Some remaining porosity must be expected here, too.

Properties

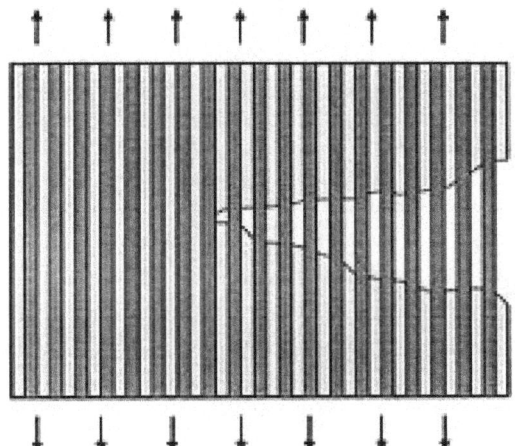

Scheme of crack bridges at the crack tip of ceramic composites.

Mechanical Properties

Basic Mechanism of Mechanical Properties

The high fracture toughness or crack resistance mentioned above is a result of the following mechanism: under load the ceramic matrix cracks, like any ceramic material, at an elongation of about 0.05%. In CMCs the embedded fibres bridge these cracks. This mechanism works only when the matrix can slide along the fibres, which means that there must be a weak bond between the fibres and matrix. A strong bond would require a very high elongation capability of the fibre bridging the crack, and would result in a brittle fracture, as with conventional ceramics. The production of CMC material with high crack resistance requires a step to weaken this bond between the fibres and matrix. This is achieved by depositing a thin layer of pyrolytic carbon or boron nitride on the fibres, which weakens the bond at the fibre/matrix interface (sometimes "interface"), leading to the fibre pull-out at crack surfaces, as shown in the SEM picture at the top of this article. In oxide-CMCs, the high porosity of the matrix is sufficient to establish the weak bond.

Properties Under Tensile and Bending Loads, Crack Resistance

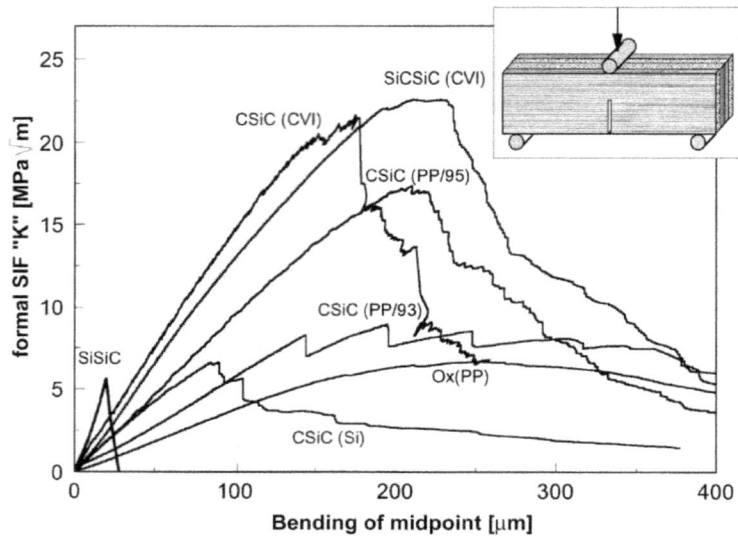

Curves of toughness measurements of various ceramic composites and SiSiC
Legend: SiSiC: conventional SiSiC, SiCSiC(CVI) and CSiC(CVI): SiC/SiC and C/SiC manufactured in CVI processes, CSiC(95) und CSiC(93): C/SiC manufactured by the LPI-method, Ox(PP): oxide ceramic composite, CSiC(Si): C/SiC manufactured via the LSI process.

The influence and quality of the fibre interface can be evaluated through mechanical properties. Measurements of the crack resistance were performed with notched specimens in so-called single-edge-notch-bend (SENB) tests. In fracture mechanics, the measured data (force, geometry and crack surface) are normalized to yield the so-called stress intensity factor (SIF), K_{Ic}. Because of the complex crack surface the real crack surface area can not be determined for CMC materials. The measurements therefore use the initial notch as the crack surface, yielding the *formal SIF* shown in the figure. This requires identical geometry for comparing different samples. The area under these curves thus gives a relative indication of the energy required to drive the crack tip through the sample (force times path length gives energy). The maxima indicate the load level necessary to propagate the crack through the sample. Compared to the sample of conventional SiSiC ceramic, two observations can be made:

- All tested CMC materials need up to several orders of magnitude more energy to propagate the crack through the material.

- The force required for crack propagation varies between different types of CMCs.

Type of material	Al_2O_3/Al_2O_3	Al_2O_3	CVI-C/SiC	LPI-C/SiC	LSI-C/SiC	SiSiC
Porosity (%)	35	<1	12	12	3	<1
Density (g/cm³)	2.1	3.9	2.1	1.9	1.9	3.1
Tensile strength (MPa)	65	250	310	250	190	200
Elongation (%)	0.12	0.1	0.75	0.5	0.35	0.05
Young's modulus (GPa)	50	400	95	65	60	395
Flexural strength (MPa)	80	450	475	500	300	400

In the table, CVI, LPI, and LSI denote the manufacturing process of the C/SiC-material. Data of the oxide CMC and SiSiC are taken from manufacturer data sheets. Tensile strength of SiSiC and Al_2O_3 were calculated from measurements of elongation to fracture and Young's modulus, since generally only bending strength data are available for those ceramics. Averaged values are given in the table, and significant differences, even within one manufacturing route, are possible.

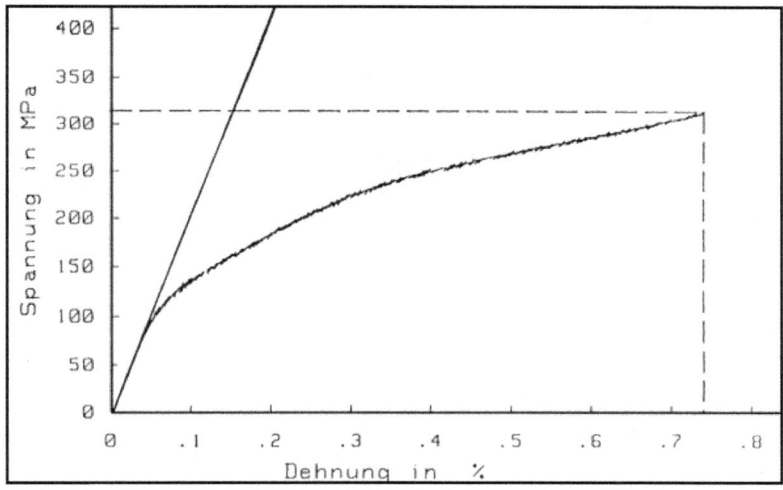

Stress-strain curve of a tensile test for CVI-SiC/SiC.

Tensile tests of CMCs usually show nonlinear stress-strain curves, which look as if the material deforms plastically. It is called *quasi-plastic*, because the effect is caused by the microcracks, which are formed and bridged with increasing load. Since the Young's modulus of the load-carrying fibres is generally lower than that of the matrix, the slope of the curve decreases with increasing load.

Curves from bending tests look similar to those of the crack resistance measurements shown above.

The following features are essential in evaluating bending and tensile data of CMCs:

- CMC materials with a low matrix content (down to zero) have a high tensile strength (close to the tensile strength of the fibre), but low bending strength.

- CMC materials with a low fibre content (down to zero) have a high bending strength (close to the strength of the monolithic ceramic), but no elongation beyond 0.05% under tensile load.

The primary quality criterion for CMCs is the crack resistance behaviour or fracture toughness.

Other Mechanical Properties

In many CMC components the fibres are arranged as 2-dimensional (2D) stacked plain or satin weave fabrics. Thus the resulting material is anisotropic or, more specifically, orthotropic. A crack between the layers is not bridged by fibres. Therefore, the interlaminar shear strength (ILS) and the strength perpendicular to the 2D fiber orientation are low for these materials. Delamination can occur easily under certain mechanical loads. Three-dimensional fibre structures can improve this situation.

Material	CVI-C/SiC	LPI-C/SiC	LSI-C/SiC	CVI-SiC/SiC
Interlaminar shear strength (MPa)	45	30	33	50
Tensile strength vertical to fabric plane (MPa)	6	4	–	7
Compressive strength vertical to fabric plane (MPa)	500	450	–	500

The compressive strengths shown in the table are lower than those of conventional ceramics, where values above 2000 MPa are common; this is a result of porosity.

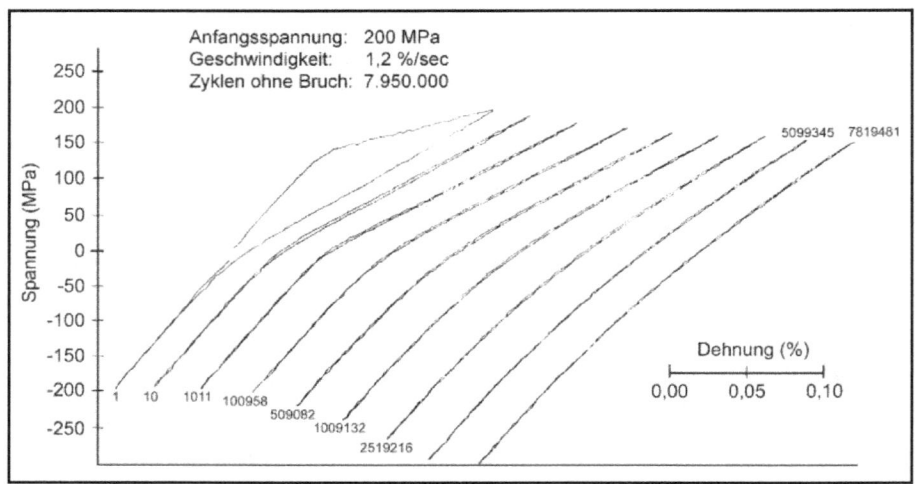

Strain controlled LCF-test for a CVI-SiC/SiC-specimen.

The composite structure allows high dynamical loads. In the so-called low-cycle-fatigue (LCF) or high-cycle-fatigue (HCF) tests the material experiences cyclic loads under tensile and compressive (LCF) or only tensile (HCF) load. The higher the initial stress the shorter the lifetime and the smaller the number of cycles to rupture. With an initial load of 80% of the strength, a SiC/SiC sample survived about 8 million cycles.

The Poisson's ratio shows an anomaly when measured perpendicular to the plane of the fabric, because interlaminar cracks increase the sample thickness.

Thermal and Electrical Properties

The thermal and electrical properties of the composite are a result of its constituents, namely fibres, matrix and pores as well as their composition. The orientation of the fibres yields anisotropic data. Oxide CMCs are very good electrical insulators, and because of their high porosity their thermal insulation is much better than that of conventional oxide ceramics.

The use of carbon fibres increases the electrical conductivity, provided the fibres contact each other and the voltage source. Silicon carbide matrix is a good thermal conductor. Electrically, it is a semiconductor, and its resistance therefore decreases with increasing temperature. Compared to (poly)crystalline SiC, the amorphous SiC fibres are relatively poor conductors of heat and electricity.

Material	CVI-C/SiC	LPI-C/SiC	LSI-C/SiC	CVI-SiC/SiC	SiSiC
Thermal conductivity (p) [W/(m·K)]	15	11	21	18	>100
Thermal conductivity (v) [W/(m·K)]	7	5	15	10	>100
Linear expansion (p) [10^{-6}·1/K]	1.3	1.2	0	2.3	4
Linear expansion (v) [10^{-6}·1/K]	3	4	3	3	4
Electrical resistivity (p) [Ω·cm]	–	–	–	–	50
Electrical resistivity (v) [Ω·cm]	0.4	–	–	5	50

Comments for the table: (p) and (v) refer to data parallel and vertical to fibre orientation of the 2D-fiber structure, respectively. LSI material has the highest thermal conductivity because of its low porosity – an advantage when using it for brake discs. These data are subject to scatter depending on details of the manufacturing processes.

Conventional ceramics are very sensitive to thermal stress because of their high Young's modulus and low elongation capability. Temperature differences and low thermal conductivity create locally different elongations, which together with the high Young's modulus generate high stress. This results in cracks, rupture and brittle failure. In CMCs, the fibres bridge the cracks, and the components show no macroscopic damage, even if the matrix has cracked locally. The application of CMCs in brake disks demonstrates the effectiveness of ceramic composite materials under extreme thermal shock conditions.

Corrosion Properties

Data on the corrosion behaviour of CMCs are scarce except for oxidation at temperatures above 1000 °C. These properties are determined by the constituents, namely the fibres and matrix. Ceramic materials in general are very stable to corrosion. The broad spectrum of manufacturing techniques with different sintering additives, mixtures, glass phases and porosities are crucial for the results of corrosion tests. Less impurities and exact stoichiometry lead to less corrosion. Amorphous structures and non-ceramic chemicals frequently used as sintering aids are starting points of corrosive attack.

Alumina

Pure alumina shows excellent corrosion resistivity against most chemicals. Amorphous glass and silica phases at the grain boundaries determine the speed of corrosion in concentrated acids and bases and result in creep at high temperatures. These characteristics limit the use of alumina. For molten metals, alumina is used only with gold and platinum.

Alumina fibres

These fibres demonstrate corrosion properties similar to alumina, but commercially available fibres are not very pure and therefore less resistant. Because of creep at temperatures above 1000 °C, there are only few applications for oxide CMCs.

Carbon

The most significant corrosion of carbon occurs in presence of oxygen above about 500 °C. It burns to form carbon dioxide and/or carbon monoxide. It also oxidises in strong oxidizing agents like concentrated nitric acid. In molten metals it dissolves and forms metal carbides. Carbon fibres do not differ from carbon in their corrosion behaviour.

Silicon carbide

Pure silicon carbide is one of the most corrosion-resistant materials. Only strong bases, oxygen above about 800 °C, and molten metals react with it to form carbides and silicides. The reaction with oxygen forms SiO_2 and CO_2, whereby a surface layer of SiO_2 slows down subsequent oxidation (*passive oxidation*). Temperatures above about 1600 °C and a low partial pressure of oxygen result in so-called *active oxidation*, in which CO, CO_2 and gaseous SiO are formed causing rapid loss of SiC. If the SiC matrix is produced other than by CVI, corrosion-resistance is not as good. This is a consequence of porosity in the amorphous LPI, and residual silicon in the LSI-matrix.

Silicon carbide fibres

Silicon carbide fibres are produced via pyrolysis of organic polymers, and therefore their corrosion properties are similar to those of the silicon carbide found in LPI-matrices. These fibres are thus more sensitive to bases and oxidizing media than pure silicon carbide.

Applications

CMC materials overcome the major disadvantages of conventional technical ceramics, namely brittle failure and low fracture toughness, and limited thermal shock resistance. Therefore, their applications are in fields requiring reliability at high-temperatures (beyond the capability of metals) and resistance to corrosion and wear. These include:

- Heat shield systems for space vehicles, which are needed during the re-entry phase, where high temperatures, thermal shock conditions and heavy vibration loads take place.

- Components for high-temperature gas turbines such as combustion chambers, stator vanes and turbine blades.

- Components for burners, flame holders, and hot gas ducts, where the use of oxide CMCs has found its way.

- Brake disks and brake system components, which experience extreme thermal shock (greater than throwing a glowing part of any material into water).

- Components for slide bearings under heavy loads requiring high corrosion and wear resistance.

In addition to the foregoing, CMCs can be used in applications, which employ conventional ceramics or in which metal components have limited lifetimes due to corrosion or high temperatures.

Developments for Applications in Space

During the re-entry phase of space vehicles, the heat shield system is exposed to temperatures above 1500 °C for a few minutes. Only ceramic materials are able to survive such conditions without significant damage, and among ceramics only CMCs can adequately handle thermal shocks. The development of CMC-based heat shield systems promises the following advantages:

- Reduced weight

- Higher load carrying capacity of the system

- Reusability for several re-entries

- Better steering during the re-entry phase with CMC flap systems

NASA-space vehicle X-38 during a test flight.

In these applications the high temperatures preclude the use of oxide fibre CMCs, because under the expected loads the creep would be too high. Amorphous silicon carbide fibres lose their strength due to re-crystallization at temperatures above 1250 °C. Therefore, carbon fibres in a silicon carbide matrix (C/SiC) are used in development programs for these applications. The European program HERMES of ESA, started in the 1980s and for financial reasons abandoned in 1992, has produced first results. Several follow-up programs focused on the development, manufacture, and qualification of nose cap, leading edges and steering flaps for the NASA space vehicle X-38.

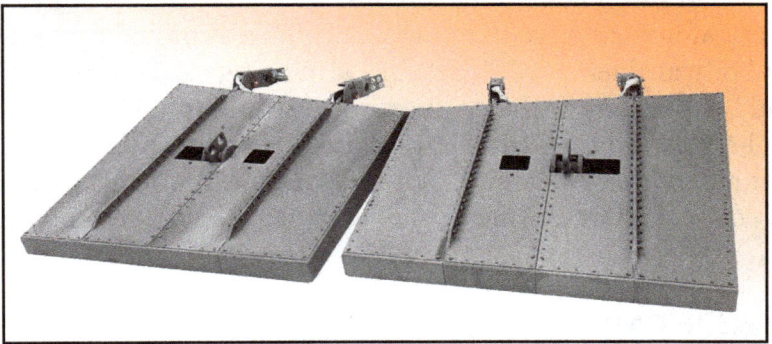

Pair of steering flaps for the NASA-space vehicle X-38. Size: 1.5×1.5×0.15 m, mass: 68 kg each, various components are mounted using more than 400 CVI-C/SiC screws and nuts.

This development program has qualified the use of C/SiC bolts and nuts, and the bearing system of the flaps. The latter were ground-tested at the DLR in Stuttgart, Germany, under expected conditions of the re-entry phase: 1600 °C, 4 tonnes load, oxygen partial pressure similar to re-entry conditions, and simultaneous bearing movements of four cycles per second. A total of five re-entry phases was simulated. Furthermore, oxidation protection systems were developed and qualified to prevent burnout of the carbon fibres. After mounting of the flaps, mechanical ground tests were performed successfully by NASA in Houston, Texas, US. The next test – a real re-entry of the un-manned vehicle X-38 – was cancelled for financial reasons. One of the space shuttles would have brought the vehicle into orbit, from where it would have returned to the Earth.

These qualifications were promising for only this application. The high-temperature load lasts only around 20 minutes per re-entry, and for reusability, only about 30 cycles would be sufficient. For industrial applications in hot gas environment, though, several hundred cycles of thermal loads and up to many thousands hours of lifetime are required.

The Intermediate Experimental Vehicle (IXV), a project initiated by ESA in 2009, is Europe's first lifting body reentry vehicle. Developed by Thales Alenia Space, the IXV is scheduled to make its first flight in 2014 on the fourth Vega mission (VV04) over the Gulf of Guinea. More than 40 European companies contributed to its construction. The thermal protection system for the underside of the vehicle, comprising the nose, leading edges and lower surface of the wing, were designed and made by Herakles using a ceramic matrix composite (CMC), carbon/silicon-carbide (C/SiC). These components will function as the vehicle's heat shield during its atmospheric reentry.

Developments for Gas Turbine Components

The use of CMCs in gas turbines would permit higher turbine inlet temperatures, which would improve turbine efficiency. Because of the complex shape of stator vanes and turbine blades, the development was first focused on the combustion chamber. In the US, a combustor made of SiC/SiC with a special SiC fiber of enhanced high-temperature stability was successfully tested for 15,000 hours. SiC oxidation was substantially reduced by the use of an oxidation protection coating consisting of several layers of oxides. The engine collaboration between General Electric and Rolls-Royce is studying the use of CMC stator vanes in the hot section of the F136 turbofan engine, an engine which failed to beat the Pratt and Whitney F-135 for use in the Joint Strike Fighter. The engine joint venture, CFM International is also considering the use of CMC parts to reduce weight in its Leap-X demonstrator engine program, which is aimed at providing next-generation turbine engines for narrow-body airliners. CMC parts are also being studied for stationary applications in both the cold and hot sections of the engines, since stresses imposed on rotating parts would require further development effort. Generally, a successful application in turbines still needs a lot of technical and cost reduction work for all high-temperature components to justify the efficiency gain. Furthermore, cost reduction for fibers, manufacturing processes and protective coatings is essential.

Application of Oxide CMC in Burner and Hot Gas Ducts

Oxygen-containing gas at temperatures above 1000 °C is rather corrosive for metal and silicon carbide components. Such components, which are not exposed to high mechanical stress, can be made of oxide CMCs, which can withstand temperatures up to 1200 °C. The gallery below shows the flame holder of a crisp bread bakery as tested after for 15,000 hours, which subsequently operated for a total of more than 20,000 hours.

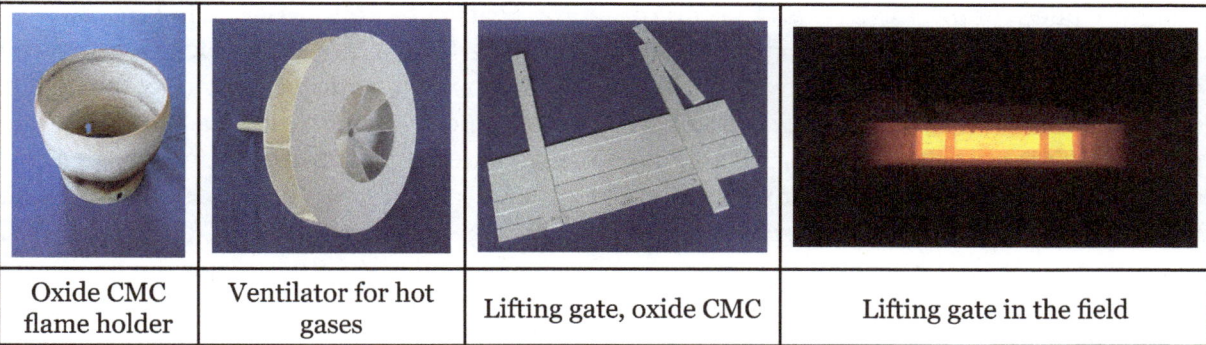

Oxide CMC flame holder	Ventilator for hot gases	Lifting gate, oxide CMC	Lifting gate in the field

Flaps and ventilators circulating hot, oxygen-containing gases can be fabricated in the same shape as their metal equivalents. The lifetime for these oxide CMC components is several times longer than for metals, which often deform. A further example is an oxide CMC lifting gate for a sintering furnace, which has survived more than 260,000 opening cycles.

Application in Brake Disk

Carbon/carbon (C/C) materials have found their way into the disk brakes of racing cars and aeroplanes, and C/SiC brake disks manufactured by the LSI process were qualified and are commercially available for luxury vehicles. The advantages of these C/SiC disks are:

- Very little wear, resulting in lifetime use for a car with a normal driving load of 300,000 km, is forecast by manufacturers.

- No fading is experienced, even under high load.

- No surface humidity effect on the friction coefficient shows up, as in C/C brake disks.

- The corrosion resistance, for example to the road salt, is much better than for metal disks.

- The disk mass is only 40% of a metal disk. This translates into less unsprung and rotating mass.

The weight reduction improves shock absorber response, road-holding comfort, agility, fuel economy, and thus driving comfort.

The SiC-matrix of LSI has a very low porosity, which protects the carbon fibres quite well. Brake disks do not experience temperatures above 500 °C for more than a few hours in their lifetime. Oxidation is therefore not a problem in this application. The reduction of manufacturing costs will decide the success of this application for middle-class cars.

Application in Slide Bearings

Conventional SiC, or sometimes the less expensive SiSiC, have been used successfully for more than 25 years in slide or journal bearings of pumps. The pumped liquid itself provides the lubricant for the bearing. Very good corrosion resistance against practically all kinds of media, and very low wear and low friction coefficients are the basis of this success. These bearings consist of a static bearing, shrink-fitted in its metallic environment, and a rotating shaft sleeve, mounted on the shaft. Under compressive stress the ceramic static bearing has a low risk of failure, but a SiC shaft sleeve does not have this situation and must therefore have a large wall thickness and/or be specially designed. In large pumps with shafts 100–350 mm in diameter, the risk of failure is higher due to the changing requirements on the pump performance – for example, load changes during operation. The introduction of SiC/SiC as a shaft sleeve material has proven to be very successful. Test rig experiments showed an almost triple specific load capability of the bearing system with a shaft sleeve made of SiC/SiC, sintered SiC as static bearing, and water at 80 °C as lubricant. The specific load capacity of a bearing is usually given in W/mm² and calculated as a product of the load (MPa), surface speed of the bearing (m/s) and friction coefficient; it is equal to the power loss of the bearing system due to friction.

Components for a ceramic slide bearing; the picture shows a sintered SiC-bearing for a hydrostatic slide bearing and a CVI-SiC/SiC-shaft sleeve shrink-fitted on metal, a system tested with liquid oxygen as lubricant.

In boiler feedwater pumps of power stations, which pump several thousand cubic meters of hot water to a level of 2000 m, and in tubular casing pumps for water works or sea water desalination plants (pumping up to 40,000 m³ to a level of around 20 m) this slide bearing concept, namely SiC/SiC shaft sleeve and SiC bearing, has been used since 1994.

This bearing system has been tested in pumps for liquid oxygen, for example in oxygen turbopumps for thrust engines of space rockets, with the following results. SiC and SiC/SiC are compatible with liquid oxygen. In an auto-ignition test according to the French standard NF 28-763, no auto-ignition was observed with powdered SiC/SiC in 20 bar pure oxygen at temperatures up to 525 °C. Tests have shown that the friction coefficient is half, and wear one fiftieth of standard metals used in this environment. A hydrostatic bearing system has survived several hours at a speed up to 10,000 revolutions per minute, various loads, and 50 cycles of start/stop transients without any significant traces of wear.

Other Applications and Developments

- Thrust control flaps for military jet engines

- Components for fusion and fission reactors

- Friction systems for various applications

- Nuclear applications

Transparent Ceramics

Many ceramic materials, both glassy and crystalline, have found use as optically transparent materials in various forms from bulk solid-state components to high surface area forms such as thin films, coatings, and fibers. Such devices have found widespread use for various applications in the electro-optical field including: optical fibers for guided lightwave transmission, optical switches, laser amplifiers and lenses, hosts for solid-state lasers and optical window materials for gas lasers, and infrared (IR) heat seeking devices for missile guidance systems and IR night vision.

Transparent spinel ($MgAl_2O_4$) ceramic is used traditionally for applications such as high-energy laser windows because of its excellent transmission in visible wavelengths and mid-wavelength infrared (0.2-5.0µm) when combined with selected materials - source: U.S. Naval Research Laboratory.

While single-crystalline ceramics may be largely defect-free (particularly within the spatial scale of the incident light wave), optical transparency in polycrystalline materials is limited by the amount of light that is scattered by their microstructural features. The amount of light scattering therefore depends on the wavelength of the incident radiation, or light.

For example, since visible light has a wavelength scale on the order of hundreds of nanometers, scattering centers will have dimensions on a similar spatial scale. Most ceramic materials, such as alumina and its compounds, are formed from fine powders, yielding a fine grained polycrystalline microstructure that is filled with scattering centers comparable to the wavelength of visible light. Thus, they are generally opaque as opposed to transparent materials. Recent nanoscale technology, however, has made possible the production of (poly)crystalline transparent ceramics such as alumina Al_2O_3, yttria alumina garnet (YAG), and neodymium-doped Nd:YAG.

Introduction

Transparent ceramics have recently acquired a high degree of interest and notoriety. Basic applications include lasers and cutting tools, transparent armor windows, night vision devices (NVD), and nose cones for heat seeking missiles. Currently available infrared (IR) transparent materials typically exhibit a trade-off between optical performance and mechanical strength. For example, sapphire (crystalline alumina) is very strong, but lacks full transparency throughout the 3–5 micrometer mid-IR range. Yttria is fully transparent from 3–5 micrometers, but lacks sufficient strength, hardness, and thermal shock resistance for high-performance aerospace applications. Not surprisingly, a combination of these two materials in the form of the yttria-alumina garnet (YAG) has proven to be one of the top performers in the field.

Synthetic sapphire - single-crystal aluminum oxide (sapphire – Al_2O_3) is a transparent ceramic.

In 1961, General Electric began selling transparent alumina Lucalox bulbs. In 1966, GE announced a ceramic "transparent as glass," called Yttralox. In 2004, Anatoly Rosenflanz and colleagues at 3M used a "flame-spray" technique to alloy aluminium oxide (or alumina) with rare-earth metal oxides in order to produce high strength glass-ceramics with good optical properties. The method avoids many of the problems encountered in conventional glass forming and may be extensible to other oxides. This goal has been readily accomplished and amply demonstrated in laboratories and research facilities worldwide using the emerging chemical processing methods encompassed by the methods of sol-gel chemistry and nanotechnology.

Many ceramic materials, both glassy and crystalline, have found use as hosts for solid-state lasers and as optical window materials for gas lasers. The first working laser was made by Theodore H. Maiman in 1960 at Hughes Research Laboratories in Malibu, who had the edge on other research teams led by Charles H. Townes at Columbia University, Arthur Schawlow at Bell Labs, and Gould at TRG (Technical Research Group). Maiman used a solid-state light-pumped synthetic ruby to produce red laser light at a wavelength of 694 nanometers (nm). Synthethic ruby lasers are still in use. Both sapphires and rubies are corundum, a crystalline form of aluminium oxide (Al2O3).

Crystals

Ruby lasers consist of single-crystal sapphire alumina (Al_2O_3) rods doped with a small concentration of chromium Cr, typically in the range of 0.05%. The end faces are highly polished with a planar and parallel configuration. Neodymium-doped YAG (Nd:YAG) has proven to be one of the best solid-state laser materials. Its indisputable dominance in a broad variety of laser applications is determined by a combination of high emission cross section with long spontaneous emission lifetime, high damage threshold, mechanical strength, thermal conductivity, and low thermal beam distortion. The fact that the Czochralski crystal growth of Nd:YAG is a matured, highly reproducible and relatively simple technological procedure adds significantly to the value of the material.

Nd:YAG lasers are used in manufacturing for engraving, etching, or marking a variety of metals and plastics. They are extensively used in manufacturing for cutting and welding steel and various alloys. For automotive applications (cutting and welding steel) the power levels are typically 1–5 kW. In addition, Nd:YAG lasers are used in ophthalmology to correct posterior capsular opacification, a condition that may occur after cataract surgery, and for peripheral iridotomy in patients with acute angle-closure glaucoma, where it has superseded surgical iridectomy. Frequency-doubled Nd:YAG lasers (wavelength 532 nm) are used for pan-retinal photocoagulation in patients with diabetic retinopathy. In oncology, Nd:YAG lasers can be used to remove skin cancers. These lasers are also used extensively in the field of cosmetic medicine for laser hair removal and the treatment of minor vascular defects such as spider veins on the face and legs. Recently used for dissecting cellulitis, a rare skin disease usually occurring on the scalp. Using hysteroscopy in the field of gynecology, the Nd:YAG laser has been used for removal of uterine septa within the inside of the uterus. In dentistry, Nd:YAG lasers are used for soft tissue surgeries in the oral cavity.

Currently, high powered Nd:glass lasers as large as a football field are used for inertial confinement fusion, nuclear weapons research, and other high energy density physics experiments.

Glasses

Glasses (non-crystalline ceramics) also are used widely as host materials for lasers. Relative to crystalline lasers, they offer improved flexibility in size and shape and may be readily manufactured as large, homogeneous, isotropic solids with excellent optical properties. The indices of refraction of glass laser hosts may be varied between approximately 1.5 and 2.0, and both the temperature coefficient of n and the strain-optical coefficient may be tailored by altering the chemical composition. Glasses have lower thermal conductivities than the alumina or YAG, however, which imposes limitations on their use in continuous and high repetition-rate applications.

The principal differences between the behavior of glass and crystalline ceramic laser host materials are associated with the greater variation in the local environment of lasing ions in amorphous solids. This leads to a broadening of the fluorescent levels in glasses. For example, the width of the Nd^{3+} emission in YAG is ~ 10 angstroms as compared to ~ 300 angstroms in typical oxide glasses. The broadened fluorescent lines in glasses make it more difficult to obtain continuous wave laser operation (CW), relative to the same lasing ions in crystalline solid laser hosts.

Several glasses are used in transparent armor, such as normal plate glass (soda-lime-silica), borosilicate glass, and fused silica. Plate glass has been the most common glass used due to its low cost. But greater requirements for the optical properties and ballistic performance have necessitated the development of new materials. Chemical or thermal treatments can increase the strength of glasses, and the controlled crystallization of certain glass compositions can produce optical quality glass-ceramics. Alstom Grid Ltd. currently produces a lithium di-silicate based glass-ceramic known as TransArm, for use in transparent armor systems. It has all the workability of an amorphous glass, but upon recrystallization it demonstrates properties similar to a crystalline ceramic. Vycor is 96% fused silica glass, which is crystal clear, lightweight and high strength. One advantage of these type of materials is that they can be produced in large sheets and other curved shapes.

Nanomaterials

It has been shown fairly recently that laser elements (amplifiers, switches, ion hosts, etc.) made from fine-grained ceramic nanomaterials—produced by the low temperature sintering of high purity nanoparticles and powders—can be produced at a relatively low cost. These components are free of internal stress or intrinsic birefringence, and allow relatively large doping levels or optimized custom-designed doping profiles. This highlights the use of ceramic nanomaterials as being particularly important for high-energy laser elements and applications.

Primary scattering centers in polycrystalline nanomaterials—made from the sintering of high purity nanoparticles and powders—include microstructural defects such as residual porosity and grain boundaries. Thus, opacity partly results from the incoherent scattering of light at internal surfaces and interfaces. In addition to porosity, most of the interfaces or internal surfaces in ceramic nanomaterials are in the form of grain boundaries which separate nanoscale regions of crystalline order. Moreover, when the size of the scattering center (or grain boundary) is reduced well below the size of the wavelength of the light being scattered, the light scattering no longer occurs to any significant extent.

In the processing of high performance ceramic nanomaterials with superior opto-mechanical properties under adverse conditions, the size of the crystalline grains is determined largely by the size of the crystalline particles present in the raw material during the synthesis or formation of the object. Thus a reduction of the original particle size well below the wavelength of visible light (~ 0.5 μm or 500 nm) eliminates much of the light scattering, resulting in a translucent or even transparent material.

Furthermore, results indicate that microscopic pores in sintered ceramic nanomaterials, mainly trapped at the junctions of microcrystalline grains, cause light to scatter and prevented true transparency. It has been observed that the total volume fraction of these nanoscale pores (both intergranular and intragranular porosity) must be less than 1% for high-quality optical transmission, i.e. the density has to be 99.99% of the theoretical crystalline density.

Lasers

Nd:YAG

For example, a 1.46 kW Nd:YAG laser has been demonstrated by Konoshima Chemical Co. in Japan. In addition, Livermore researchers realized that these fine-grained ceramic nanomaterials might greatly benefit high-powered lasers used in the National Ignition Facility (NIF) Programs Directorate. In particular, a Livermore research team began to acquire advanced transparent nanomaterials from Konoshima to determine if they could meet the optical requirements needed for Livermore's Solid-State Heat Capacity Laser (SSHCL). Livermore researchers have also been testing applications of these materials for applications such as advanced drivers for laser-driven fusion power plants.

Assisted by several workers from the NIF, the Livermore team has produced 15 mm diameter samples of transparent Nd:YAG from nanoscale particles and powders, and determined the most important parameters affecting their quality. In these objects, the team largely followed the Japanese production and processing methodologies, and used an in house furnace to vacuum sinter the nanopowders. All specimens were then sent out for hot isostatic pressing (HIP). Finally, the components were returned to Livermore for coating and testing, with results indicating exceptional optical quality and properties.

One Japanese/East Indian consortium has focused specifically on the spectroscopic and stimulated emission characteristics of Nd^{3+} in transparent YAG nanomaterials for laser applications. Their materials were synthesized using vacuum sintering techniques. The spectroscopic studies suggest overall improvement in absorption and emission and reduction in scattering loss. Scanning electron microscope and transmission electron microscope observations revealed an excellent optical quality with low pore volume and narrow grain boundary width. Fluorescence and Raman measurements reveal that the Nd^{3+} doped YAG nanomaterial is comparable in quality to its single-crystal counterpart in both its radiative and non-radiative properties. Individual Stark levels are obtained from the absorption and fluorescence spectra and are analyzed in oredr to identify the stimulated emission channels possible in the material. Laser performance studies favor the use of high dopant concentration in the design of an efficient microchip laser. With 4 at% dopant, the group obtained a slope efficiency of 40%. High-power laser experiments yield an optical-to-optical conversion efficiency of 30% for Nd (0.6 at%) YAG nanomaterial as compared to 34% for an Nd (0.6 at%) YAG single crystal. Optical gain measurements conducted in these materials also show values comparable to single crystal, supporting the contention that these materials could be suitable substitutes to single crystals in solid-state laser applications.

Yttria, Y_2O_3

The initial work in developing transparent yttrium oxide nanomaterials was carried out by General Electric in the 1960s.

In 1966, a transparent ceramic, Yttralox, was invented by Dr. Richard C. Anderson at the General Electric Research Laboratory, with further work at GE's Metallurgy and Ceramics Laboratory by Drs. Paul J. Jorgensen, Joseph H. Rosolowski, and Douglas St. Pierre. Yttralox is "transparent as glass," has a melting point twice as high, and transmits frequencies in the near infrared band as well as visible light.

IR 100 Award, Yttralox, 1967.

Gemstones of Yttralox transparent ceramic.

Richard C. Anderson holding a sample of Yttralox.

Further development of yttrium ceramic nanomaterials was carried out by General Electric in the 1970s in Schenectady and Cleveland, motivated by lighting and ceramic laser applications. Yttralox, transparent yttrium oxide Y_2O_3 containing ~ 10% thorium oxide (ThO_2) was fabricated by Greskovich and Woods. The additive served to control grain growth during densification, so that porosity remained on grain boundaries and not trapped inside grains where it would be quite difficult to eliminate during the initial stages of sintering. Typically, as polycrystalline ceramics densify during heat treatment, grains grow in size while the remaining porosity decreases both in volume fraction and in size. Optically transparent ceramics must be virtually pore-free.

GE's transparent Yttralox was followed by GTE's lanthana-doped yttria with similar level of additive. Both of these materials required extended firing times at temperatures above 2000 °C. La_2O_3 – doped Y_2O_3 is of interest for infrared (IR) applications because it is one of the longest wavelength transmitting oxides. It is refractory with a melting point of 2430 °C and has a moderate coefficient of thermal expansion coefficient. The thermal shock and erosion resistance is considered to be intermediate among the oxides, but outstanding compared to non-oxide IR transmitting materials. A major consideration is the low emissivity of yttria, which limits background radiation upon heating. It is also known that the phonon edge gradually moves to shorter wavelengths as a material is heated.

In addition, ytrria itself, Y_2O_3 has been clearly identified as a prospective solid-state laser material. In particular, lasers with ytterbium as dopant allow the efficient operation both in cw operation and in pulsed regimes.

At high concentration of excitations (of order of 1%) and poor cooling, the quenching of emission at laser frequency and avalanche broadband emission takes place.

Future

The Livermore team is also exploring new ways to chemically synthesize the initial nanopowders. Borrowing on expertise developed in CMS over the past 5 years, the team is synthesizing nano-powders based on sol-gel processing, and then sintering them accordingly in order to obtain the solid-state laser components. Another technique being tested utilizes a combustion process in order to generate the powders by burning an organic solid containing yttrium, aluminum, and neodymium. The smoke is then collected, which consists of spherical nanoparticles.

The Livermore team is also exploring new forming techniques (e.g. extrusion molding) which have the capacity to create more diverse, and possibly more complicated, shapes. These include shells and tubes for improved coupling to the pump light and for more efficient heat transfer. In addition, different materials can be co-extruded and then sintered into a monolithic transparent solid. An amplifier slab can formed so that part of the structure acts in guided lightwave transmission in order to focus pump light from laser diodes into regions with a high concentration of dopant ions near the slab center.

In general, nanomaterials promise to greatly expand the availability of low-cost, high-end laser components in much larger sizes than would be possible with traditional single crystalline ceramics. Many classes of laser designs could benefit from nanomaterial-based laser structures such as amplifies with built-in edge claddings. Nanomaterials could also provide more robust and compact designs for high-peak power, fusion-class lasers for stockpile stewardship, as well as high-average-power lasers for global theater ICBM missile defense systems (e.g. Strategic Defense Initiative SDI, or more recently the Missile Defense Agency.

Night Vision

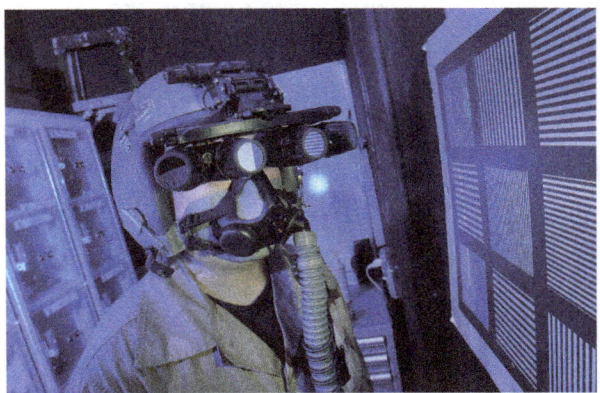

Panoramic Night Vision Goggles in testing.

A night vision device (NVD) is an optical instrument that allows images to be produced in levels of light approaching total darkness. They are most often used by the military and law enforcement agencies, but are available to civilian users. Night vision devices were first used in World War II, and came into wide use during the Vietnam War. The technology has evolved greatly since their introduction, leading to several "generations" of night vision equipment with performance increasing and price decreasing. The United States Air Force is experimenting with Panoramic Night Vision Goggles (PNVGs) which double the user's field of view to approximately 95 degrees by using four 16 mm image intensifiers tubes, rather than the more standard two 18 mm tubes.

Thermal images are visual displays of the amount of infrared (IR) energy emitted, transmitted, and reflected by an object. Because there are multiple sources of the infrared energy, it is difficult to get an accurate temperature of an object using this method. A thermal imaging camera is capable of performing algorithms to interpret that data and build an image. Although the image shows the viewer an approximation of the temperature at which the object is operating, the camera is using multiple sources of data based on the areas surrounding the object to determine that value rather than detecting the temperature.

Night vision infrared devices image in the near-infrared, just beyond the visual spectrum, and can see emitted or reflected near-infrared in complete visual darkness. All objects above the absolute zero temperature (0 K) emit infrared radiation. Hence, an excellent way to measure thermal variations is to use an infrared vision device, usually a focal plane array (FPA) infrared camera capable of detecting radiation in the mid (3 to 5 μm) and long (7 to 14 μm) wave infrared bands, denoted as MWIR and LWIR, corresponding to two of the high transmittance infrared windows. Abnormal temperature profiles at the surface of an object are an indication of a potential problem. Infrared thermography, thermal imaging, and thermal video, are examples of infrared imaging science. Thermal imaging cameras detect radiation in the infrared range of the electromagnetic spectrum (roughly 900–14,000 nanometers or 0.9–14 μm) and produce images of that radiation, called *thermograms*.

Since infrared radiation is emitted by all objects near room temperature, according to the black body radiation law, thermography makes it possible to see one's environment with or without visible illumination. The amount of radiation emitted by an object increases with temperature. Therefore, thermography allows one to see variations in temperature. When viewed through a thermal imaging camera, warm objects stand out well against cooler backgrounds; humans and other warm-blooded animals become easily visible against the environment, day or night. As a result, thermography is particularly useful to the military and to security services.

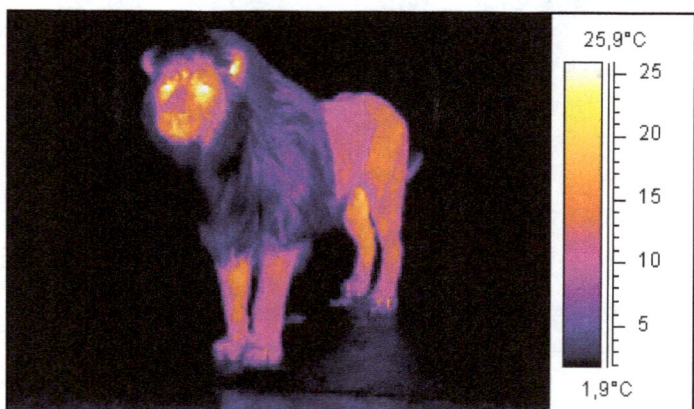

Thermogram of a lion.

Thermography

In thermographic imaging, infrared radiation with wavelengths between 8–13 micrometers strikes the detector material, heating it, and thus changing its electrical resistance. This resistance change is measured and processed into temperatures which can be used to create an image. Unlike other types of infrared detecting equipment, microbolometers utilizing a transparent ceramic detector do not require cooling. Thus, a microbolometer is essentially an uncooled thermal sensor.

The material used in the detector must demonstrate large changes in resistance as a result of minute changes in temperature. As the material is heated, due to the incoming infrared radiation, the resistance of the material decreases. This is related to the material's temperature coefficient of resistance (TCR) specifically its negative temperature coefficient. Industry currently manufactures microbolometers that contain materials with TCRs near −2%.

VO_2 and V_2O_5

The most commonly used ceramic material in IR radiation microbolometers is vanadium oxide. The various crystalline forms of vanadium oxide include both VO_2 and V_2O_5. Deposition at high temperatures and performing post-annealing allows for the production of thin films of these crystalline compounds with superior properties, which may be easily integrated into the fabrication process. VO_2 has low resistance but undergoes a metal-insulator phase change near 67 °C and also has a lower TCR value. On the other hand, V_2O_5 exhibits high resistance and also high TCR.

Other IR transparent ceramic materials that have been investigated include doped forms of CuO, MnO and SiO.

Missiles

AIM-9 Sidewinder	
Place of origin	United States

Many ceramic nanomaterials of interest for transparent armor solutions are also used for electromagnetic (EM) windows. These applications include radomes, IR domes, sensor protection, and multi-spectral windows. Optical properties of the materials used for these applications are critical, as the transmission window and related cut-offs (UV − IR) control the spectral bandwidth over which the window is operational. Not only must these materials possess abrasion resistance and strength properties common of most armor applications, but due to the extreme temperatures associated with the environment of military aircraft and missiles, they must also possess excellent thermal stability.

Thermal radiation is electromagnetic radiation emitted from the surface of an object which is due to the object's temperature. Infrared homing refers to a passive missile guidance system which uses the emission from a target of electromagnetic radiation in the infrared part of the spectrum to track it. Missiles that use infrared seeking are often referred to as "heat-seekers", since infrared is just below the visible spectrum of light in frequency and is radiated strongly by hot bodies. Many objects such as people, vehicle engines and aircraft generate and retain heat, and as such, are especially visible in the infrared wavelengths of light compared to objects in the background.

Sapphire

The current material of choice for high-speed infrared-guided missile domes is single-crystal sapphire. The optical transmission of sapphire does not extend to cover the entire mid-infrared range (3–5 µm), but starts to drop off at wavelengths greater than approximately 4.5 µm at room temperature. While the strength of sapphire is better than that of other available mid-range infrared dome materials at room temperature, it weakens above ~600 °C.

Limitations to larger area sapphires are often business related, in that larger induction furnaces and costly tooling dies are necessary in order to exceed current fabrication limits. However, as an industry, sapphire producers have remained competitive in the face of coating-hardened glass and new ceramic nanomaterials, and still managed to offer high performance and an expanded market.

Yttria, Y_2O_3

Alternative materials, such as yttrium oxide, offer better optical performance, but inferior mechanical durability. Future high-speed infrared-guided missiles will require new domes that are substantially more durable than those in use today, while still retaining maximum transparency across a wide wavelength range. A long-standing trade-off exists between optical bandpass and mechanical durability within the current collection of single-phase infrared transmitting materials, forcing missile designers to compromise on system performance. Optical nanocomposites may present the opportunity to engineer new materials that overcome this traditional compromise.

The first full scale missile domes of transparent yttria manufactured from nanoscale ceramic powders were developed in the 1980s under Navy funding. Raytheon perfected and characterized its undoped polycrystalline yttria, while lanthana-doped yttria was similarly developed by GTE Labs. The two versions had comparable IR transmittance, fracture toughness, and thermal expansion, while the undoped version exhibited twice the value of thermal conductivity.

Renewed interest in yttria windows and domes has prompted efforts to enhance mechanical properties by using nanoscale materials with submicrometer or nanosized grains. In one study, three vendors were selected to provide nanoscale powders for testing and evaluation, and they were compared to a conventional (5 µm) yttria powder previously used to prepare transparent yttria. While all of the nanopowders evaluated had impurity levels that were too high to allow processing to full transparency, 2 of them were processed to theoretical density and moderate transparency. Samples were sintered to a closed pore state at temperatures as low as 1400 C.

After the relatively short sintering period, the component is placed in a hot isostatic press (HIP) and processed for 3 – 10 hours at ~ 30 kpsi(~200 MPa) at a temperature similar to that of the initial sintering. The applied isostatic pressure provides additional driving force for densification by substantially increasing the atomic diffusion coefficients, which promotes additional viscous flow at or near grain boundaries and intergranular pores. Using this method, transparent yttria nanomaterials were produced at lower temperatures, shorter total firing times, and without extra additives which tend to reduce the thermal conductivity.

Recently, a newer method has been devleoped by Mouzon, which relies on the methods of glass-encapsulation, combined with vacuum sintering at 1600 °C followed by hot isostatic pressing (HIP) at 1500 °C of a highly agglomerated commercial powder. The use of evacuated glass capsules to perform HIP treatment allowed samples that showed open porosity after vacuum sintering to be sintered to transparency. The sintering response of the investigated powder was studied by careful microstructural observations using scanning electron microscopy and optical microscopy both in reflection and transmission. The key to this method is to keep porosity intergranular during pre-sintering, so that it can be removed subsequently by HIP treatment. It was found that agglomerates of closely packed particles are helpful to reach that purpose, since they densify fully and leave only intergranular porosity.

Composites

Prior to the work done at Raytheon, optical properties in nanocomposite ceramic materials had received little attention. Their studies clearly demonstrated near theoretical transmission in nanocomposite optical ceramics for the first time. The yttria/magnesia binary system is an ideal model system for nanocomposite formation. There is limited solid solubility in either one of the constituent phases, permitting a wide range of compositions to be investigated and compared to each other. According to the phase diagram, bi-phase mixtures are stable for all temperatures below ~ 2100 °C. In addition, neither yttria nor magnesia shows any absorption in the $3 - 5$ μm mid-range IR portion of the EM spectrum.

In optical nanocomposites, two or more interpenetrating phases are mixed in a sub-micrometer grain sized, fully dense body. Infrared light scattering can be minimized (or even eliminated) in the material as long as the grain size of the individual phases is significantly smaller than infrared wavelengths. Experimental data suggests that limiting the grain size of the nanocomposite to approximately 1/15th of the wavelength of light is sufficient to limit scattering.

Nanocomposites of yttria and magnesia have been produced with a grain size of approximately 200 nm. These materials have yielded good transmission in the 3–5 μm range and strengths higher than that for single-phase individual constituents. Enhancement of mechanical properties in nanocomposite ceramic materials has been extensively studied. Significant increases in strength (2–5 times), toughness (1–4 times), and creep resistance have been observed in systems including SiC/Al_2O_3, SiC/Si_3N_4, SiC/MgO, and Al_2O_3/ZrO_2.

The strengthening mechanisms observed vary depending on the material system, and there does not appear to be any general consensus regarding strengthening mechanisms, even within a given system. In the SiC/Al_2O_3 system, for example, it is widely known and accepted that the addition of SiC particles to the Al_2O_3 matrix results in a change of failure mechanism from intergranular (between grains) to intragranular (within grains) fracture. The explanations for improved strength include:

- A simple reduction of processing flaw concentration during nanocomposite fabrication.

- Reduction of the critical flaw size in the material—resulting in increased strength as predicted by the Hall-Petch relation).

- Crack deflection at nanophase particels due to residual thermal stresses introduced upon cooling form processing temperatures.

- Microcracking along stress-induced dislocations in the matrix material.

Armor

There is an increasing need in the military sector for high-strength, robust materials which have the capability to transmit light around the visible (0.4–0.7 micrometers) and mid-infrared (1–5 micrometers) regions of the spectrum. These materials are needed for applications requiring transparent armor. Transparent armor is a material or system of materials designed to be optically transparent, yet protect from fragmentation or ballistic impacts. The primary requirement for a transparent armor system is to not only defeat the designated threat but also provide a multi-hit capability with minimized distortion of surrounding areas. Transparent armor windows must also be compatible with night vision equipment. New materials that are thinner, lightweight, and offer better ballistic performance are being sought.

Existing transparent armor systems typically have many layers, separated by polymer (e.g. polycarbonate) interlayers. The polymer interlayer is used to mitigate the stresses from thermal expansion mismatches, as well as to stop crack propagation from ceramic to polymer. The polycarbonate is also currently used in applications such as visors, face shields and laser protection goggles. The search for lighter materials has also led to investigations into other polymeric materials such as transparent nylons, polyurethane, and acrylics. The optical properties and durability of transparent plastics limit their use in armor applications. Investigations carried out in the 1970s had shown promise for the use of polyurethane as armor material, but the optical properties were not adequate for transparent armor applications.

Several glasses are utilized in transparent armor, such as normal plate glass (soda-lime-silica), borosilicate glasses, and fused silica. Plate glass has been the most common glass used due to its low cost, but greater requirements for the optical properties and ballistic performance have generated the need for new materials. Chemical or thermal treatments can increase the strength of glasses, and the controlled crystallization of certain glass systems can produce transparent glass-ceramics. Alstom Grid Research & Technology (Stafford, UK), produced a lithium disilicate based glass-ceramic known as TransArm, for use in transparent armor systems with continuous production yielding vehicle windscreen sized pieces (and larger). The inherent advantages of glasses and glass-ceramics include having lower cost than most other ceramic materials, the ability to be produced in curved shapes, and the ability to be formed into large sheets.

Transparent crystalline ceramics are used to defeat advanced threats. Three major transparent candidates currently exist: aluminum oxynitride (AlON), magnesium aluminate spinel (spinel), and single crystal aluminum oxide (sapphire). Aluminum oxynitride spinel ($Al_{23}O_{27}N_5$), one of the leading candidates for transparent armor, is produced by Surmet Corporation as AlON and marketed under the trade name ALON. The incorporation of nitrogen into an aluminum oxide stabilizes a spinel phase, which due to its cubic crystal structure, is an isotropic material that can be produced as a transparent polycrystalline material. Polycrystalline materials can be produced in complex geometries using conventional ceramic forming techniques such as pressing, (hot) isostatic pressing, and slip casting.

Aluminium Oxynitride Spinel

Aluminium oxynitride spinel ($Al_{23}O_{27}N_5$), abbreviated as AlON, is one of the leading candidates for transparent armor. It is produced by the Surmet Corporation under the trademark ALON. The incorporation of nitrogen into aluminium oxide stabilizes a crystalline spinel phase, which due to its cubic crystal structure and unit cell, is an isotropic material which can be produced as transparent ceramic nanomaterial. Thus, fine-grained polycrystalline nanomaterials can be produced and formed into complex geometries using conventional ceramic forming techniques such as hot pressing and slip casting.

The Surmet Corporation has acquired Raytheon's ALON business and is currently building a market for this technology in the area of Transparent Armor, Sensor windows, Reconnaissance windows and IR Optics such as Lenses and Domes and as an alternative to quartz and sapphire in the semiconductor market. The AlON based transparent armor has been tested to stop multi-hit threats including of 30calAPM2 rounds and 50calAPM2 rounds successfully. The high hardness of AlON provides a scratch resistance which exceeds even the most durable coatings for glass scanner windows, such as those used in supermarkets. Surmet has successfully produced a 15"x18" curved AlON window and is currently attempting to scale up the technology and reduce the cost. In addition, the U.S. Army and U.S. Air Force are both seeking development into next generation applications.

Spinel

Magnesium aluminate spinel ($MgAl_2O_4$) is a transparent ceramic with a cubic crystal structure with an excellent optical transmission from 0.2 to 5.5 micrometers in its polycrystalline form. Optical quality transparent spinel has been produced by sinter/HIP, hot pressing, and hot press/HIP operations, and it has been shown that the use of a hot isostatic press can improve its optical and physical properties.

Spinel offers some processing advantages over AlON, such as the fact that spinel powder is available from commercial manufacturers while AlON powders are proprietary to Raytheon. It is also capable of being processed at much lower temperatures than AlON and has been shown to possess superior optical properties within the infrared (IR) region. The improved optical characteristics make spinel attractive in sensor applications where effective communication is impacted by the protective missile dome's absorption characteristics.

Spinel shows promise for many applications, but is currently not available in bulk form from any manufacturer, although efforts to commercialize spinel are underway. The spinel products business is being pursued by two key U.S. manufacturers: "Technology Assessment and Transfer" and the "Surmet Corporation".

An extensive NRL review of the literature has indicated clearly that attempts to make high-quality spinel have failed to date because the densification dynamics of spinel are poorly understood. They have conducted extensive research into the dynamics involved during the densification of spinel. Their research has shown that LiF, although necessary, also has extremely adverse effects during the final stages of densification. Additionally, its distribution in the precursor spinel powders is of critical importance.

Traditional bulk mixing processes used to mix LiF sintering aid into a powder leave fairly inhomogeneous distribution of Lif that must be homogenized by extended heat treatment times at elevated temperatures. The homogenizing temperature for Lif/Spinel occurs at the temperature of fast reaction between the LiF and the Al_2O_3. In order to avoid this detrimental reaction, they have developed a new process that uniformly coats the spinel particles with the sintering aid. This allows them to reduce the amount of Lif necessary for densification and to rapidly heat through the temperature of maximum reactivity. These developments have allowed NRL to fabricate $MgAl_2O_4$ spinel to high transparency with extremely high reproducibility that should enable military as well as commercial use of spinel.

Sapphire

Single-crystal aluminum oxide (sapphire – Al_2O_3) is a transparent ceramic. Sapphire's crystal structure is rhombohedral and thus its properties are anisotropic, varying with crystallographic orientation. Transparent alumina is currently one of the most mature transparent ceramics from a production and application perspective, and is available from several manufacturers. But the cost is high due to the processing temperature involved, as well as machining costs to cut parts out of single crystal boules. It also has a very high mechanical strength – but that is dependent on the surface finish.

The high level of maturity of sapphire from a production and application standpoint can be attributed to two areas of business: electromagnetic spectrum windows for missiles and domes, and electronic/semiconductor industries and applications.

There are current programs to scale-up sapphire grown by the heat exchanger method or edge defined film-fed growth (EFG) processes. Its maturity stems from its use as windows and in semiconductor industry. Crystal Systems Inc. which uses single crystal growth techniques, is currently scaling their sapphire boules to 13-inch (330 mm) diameter and larger. Another producer, the Saint-Gobain Group produces transparent sapphire using an edge, defined growth technique. Sapphire grown by this technique produces an optically inferior material to that which is grown via single crystal techniques, but is much less expensive, and retains much of the hardness, transmission, and scratch-resistant characteristics. Saint-Gobain is currently capable of producing 0.43" thick (as grown) sapphire, in 12" × 18.5" sheets, as well as thick, single-curved sheets. The U.S. Army Research Laboratory is currently investigating use of this material in a laminate design for transparent armor systems. The Saint Gobain Group have commercialized the capability to meet flight requirements on the F-35 Joint Strike Fighter and F-22 Raptor next generation fighter aircraft.

Composites

Future high-speed infrared-guided missiles will require new dome materials that are substantially more durable than those in use today, while retaining maximum transparency across the entire operational spectrum or bandwidth. A long-standing compromise exists between optical bandpass and mechanical durability within the current group of single-phase (crystalline or glassy) IR transmitting ceramic materials, forcing missile designers to accept substandard overall system performance. Optical nanocomposites may provide the opportunity to engineer new materials that may overcome these traditional limitations.

For example, transparent ceramic armor consisting of a lightweight composite has been formed by utilizing a face plate of transparent alumina Al_2O_3 (or magnesia MgO) with a back-up plate of transparent plastic. The two plates (bonded together with a transparent adhesive) afford complete ballistic protection against 0.30 AP M2 projectiles at 0° obliquity with a muzzle velocity of 2,770 ft (840 m) per second. Another transparent composite armor provided complete protection for small arms projectiles up to and including caliber .50 AP M2 projectiles consisting of two or more layers of transparent ceramic material.

Nanocomposites of yttria and magnesia have been produced with an average grain size of ~200 nm. These materials have exhibited near theoretical transmission in the $3 - 5$ μm IR band. Additionally, such composites have yielded higher strengths than those observed for single phase solid-state components. Despite a lack of agreement regarding mechanism of failure, it is widely accepted that nanocomposite ceramic materials can and do offer improved mechanical properties over those of single phase materials or nanomaterials of uniform chemical composition.

It should also be noted here that nanocomposite ceramic materials also offer interesting mechanical properties not achievable in other materials, such as superplastic flow and metal-like machinability. It is anticipated that further development will result in high strength, high transparency nanomaterials which are suitable for application as next generation armor.

References

- N. Miriyala; J. Kimmel; J. Price; H. Eaton; G. Linsey; E. Sun (2002). "The evaluation of CFCC Liners After Field Testing in a Gas Turbine – III" (PDF): 109–118. doi:10.1115/GT2002-30585. ISBN 0-7918-3609-6

- N.P. Bansal, J.Lamon (ed.): "Ceramic Matrix Composites: Materials, Modeling and Technology". Wiley, Hoboken, NJ 2015. ISBN 978-1-118-23116-6, p. 609

- Advances in Ceramic Armor IV. Part I: Transparent Glasses and Ceramics], Ceramic Engineering and Science Proceedings, Vol. 29 (Wiley, American Ceramic Society, 2008) ISBN 0-470-34497-0

- W. Krenkel, R. Renz, CMCs for Friction Applications, in Ceramic Matrix Composites, W. Krenkel editor, Wiley-VCH, 2008. ISBN 978-3-527-31361-7, p. 396

- Gray, Theodore (2012-04-03). The Elements: A Visual Exploration of Every Known Atom in the Universe. Black Dog & Leventhal Publishers. ISBN 9781579128951. Retrieved 26, May 2020

- P. Boullon; G. Habarou; P.C. Spriet; J.L. Lecordix; G.C. Ojard; G.D. Linsey; D.T. Feindel (2002). "Volume 4: Turbo Expo 2002, Parts A and B": 15–21. doi:10.1115/GT2002-30458. ISBN 0-7918-3609-6

- K.L. More; P.F. Tortorelli; L.R. Walker; J.B. Kimmel; N. Miriyala; J.R. Price; H.E. Eaton; E. Y. Sun; G.D. Linsey (2002). "Volume 4: Turbo Expo 2002, Parts A and B" (PDF): 155–162. doi:10.1115/GT2002-30630. ISBN 0-7918-3609-6

- Antram, Nicholas; Morrice, Richard (2008). Brighton and Hove. Pevsner Architectural Guides. London: Yale University Press. ISBN 978-0-300-12661-7

- Peng, Hong (2004). Spark Plasma Sintering of Si3N4-based Ceramics: Sintering mechanism-Tailoring microstructure-Evaluating properties (PhD thesis). Stockholm University. ISBN 978-91-7265-834-9

- Nishi, Yoshio; Doering, Robert (2000). Handbook of semiconductor manufacturing technology. CRC Press. pp. 324–325. ISBN 0-8247-8783-8

Ceramic Forming and Casting

Ceramic forming techniques include the various ways used in the formation of ceramics. Some of the techniques of ceramic forming are freeze-casting, freeze gelation and sintering. Freeze-casting is a process that produces materials with complex and three-dimensional pore structures. The topics elaborated in this chapter will help in gaining a better perspective about these ceramic forming techniques.

Ceramic Forming

Ceramic forming techniques are ways of forming ceramics, which are used to make everyday tableware from teapots, to engineering ceramics such as computer parts. Pottery techniques include the potter's wheel, slipcasting, and many others.

Methods for forming powders of ceramic raw materials into complex shapes are desirable in many areas of technology. For example, such methods are required for producing advanced, high-temperature structural parts such as heat engine components, recuperators and the like from powders of ceramic raw materials. Typical parts produced with this production operation include impellers made from stainless steel, bronze, complex cutting tools, plastic mould tooling, and others. Typical materials used are: wood, metal, water, plaster, epoxy and STLs, silica, and zirconia.

This production operation is well known for providing tools with dimensional stability, surface quality, density and uniformity. For instance, on the slip casting process the cast part is of high concentration of raw materials with little additive, this improves uniformity. But also, the plaster mould draws water from the poured slip to compact and form the casting at the mould surface. This forms a dense cast.

Slip Casting

There are many forming techniques to make ceramics, but one example is slipcasting. This is where slip, liquid clay, is poured into a plaster mould. The water in the slip is drawn out of the slip, leaving an inside layer of solid clay. When this is thick enough, the excess slip can be removed from the mould. When dry, the solid clay can then also be removed. The slip used in slip casting is often liquified with a substance that reduces the need for additional water to soften the slip; this prevents excessive shrinkage which occurs when a piece containing a lot of water dries.

Slip-casting methods provide superior surface quality, density and uniformity in casting high-purity ceramic raw materials over other ceramic casting techniques, such as hydraulic casting, since the cast part is a higher concentration of ceramic raw materials with little additives. A slip is a suspension of fine raw materials powder in a liquid such as water or alcohol with small amounts of secondary materials such as dispersants, surfactants and binders. Pottery slipcasting techniques employ a plaster block or flask mould. The plaster mould draws water from the poured slip to compact and form the casting at the mould surface. This forms a dense cast removing deleterious air gaps and minimizing shrinkage in the final sintering process.

Ceramic Shell Casting

Ceramic shell casting techniques using silica, zirconia and other refractory materials are currently used by the metal parts industry for 'net casting', forming precision shell moulds for molten metal casting. The technique involves a successive wet dipping and dry powder coating or stucco to build up the mould shell layer. The shell casting method in general is known for dimensional stability and is used in many net-casting processes for aerospace and other industries in molten metal casting. Automated facilities use multiple wax patterns on trees, large slurry mixers and fluidic powder beds for automated dipping.

Technical Ceramics

When forming technical ceramic materials from dry powders prepared for processing, the method of forming into the shape required depends upon the method of material preparation and size and shape of the part to be formed. Materials prepared for dry powder forming are most commonly formed by "dry" pressing in mechanical or hydraulic powder compacting presses selected for the necessary force and powder fill depth. Dry powder is automatically discharged into the non-flexible steel or tungsten carbide insert in the die and punches then compact the powder to the shape of the die. If the part is to be large and unable to have pressure transmit suitably for a uniform pressed density then isostatic pressing may be used. When iso-statically pressed the powder takes the shape of a flexible membrane acting as the mould, forming the shape and size of the pressed powder. Isostatic presses can be either high speed, high output type of automatic presses for such parts as ceramic insulators for spark plugs or sand blast nozzles, or slower operating "wet bag" presses that are much more manual in operation but suitable particularly for large machinable blanks or blanks that will be cut or otherwise formed in secondary operations to the final shape.

If technical ceramic parts are needed where the length to diameter ratio is very large, extrusion may be used. There are two types of ceramic extruders one being piston type with hydraulic force pushing a ram that in turn is pushing the ceramic through the loaded material cylinder to and through the die which forms the extrudate. The second type of extruder is a screw, or auger, type where a screw turns forcing the material to and through the die which again shapes the part. In both types of extrusion the raw material must be plasticized to allow and induce the flow of the material in the process.

Complex technical ceramic parts are commonly formed using either the injection moulding process or "hot wax moulding." Both rely on heat sensitive plasticizers to allow material flow into a die. The part is then quickly cooled for removal from the die. Ceramic injection moulding is much like plastic injection moulding using various polymers for plasticizing. Hot wax moulding largely uses paraffin wax.

Other Techniques

There are also several traditional techniques of handbuilding, such as pinching, soft slab, hard slab, and coil construction.

Other techniques involve threading animal or artificial wool fiber through paperclay slip, to build up layers of material. The result can be wrapped over forms or cut, dried and later joined with liquid and soft paperclay.

When forming very thin sheets of ceramic material, "tape casting" is commonly used. This involves pouring the slip (which contains a polymer "binder" to give it strength) onto a moving carrier belt, and then passing it under a stationary "doctor blade" to adjust the thickness. The moving slip is then air dried, and the "tape" thus formed is peeled off the carrier belt, cut into rectangular shapes, and processed further. As many as 100 tape layers, alternating with conductive metal powder layers, can be stacked up. These are then sintered ("fired") to remove the polymer and thus make "multilayer" capacitors, sensors, etc. According to D. W. Richerson of the American Ceramic Society, more than a billion of such capacitors are manufactured every day. (About 100 are in a typical cellular telephone, and about a thousand in a typical automobile.)

Gel casting is another technique used to create engineering ceramics.

Freeze-casting

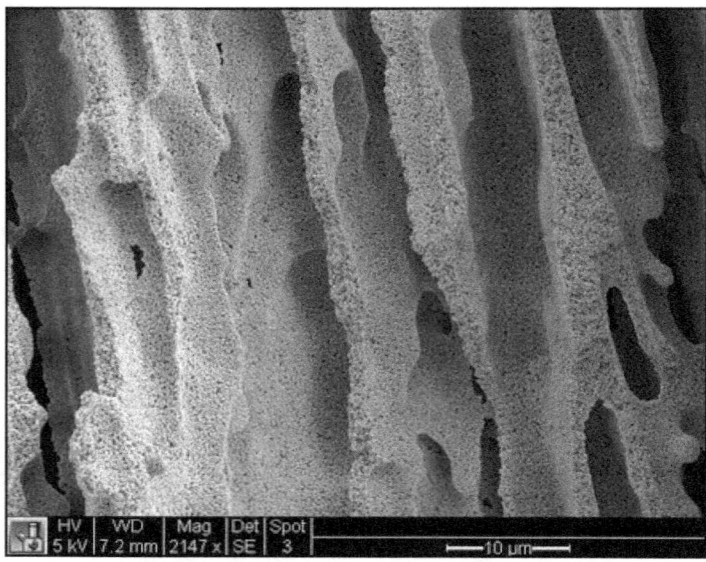

Freeze-Cast alumina that has been partially sintered. The freezing direction in the image is up.

Freeze-casting is a technique that exploits the highly anisotropic solidification behavior of a solvent (generally water) in a well-dispersed slurry to template controllably a directionally porous ceramic. By subjecting an aqueous slurry to a directional temperature gradient, ice crystals will nucleate on one side of the slurry and grow along the temperature gradient. The ice crystals will redistribute the suspended ceramic particles as they grow within the slurry, effectively templating the ceramic.

Once solidification has ended, the frozen, templated ceramic is placed into a freeze-dryer to remove the ice crystals. The resulting green body contains anisotropic macropores in an replica of the sublimated ice crystals and micropores found between the ceramic particles in the walls. This structure is often sintered to consolidate the particulate walls and provide strength to the porous material. The porosity left by the sublimation of solvent crystals is typically between 2 - 200 μm.

Overview

The first observation of cellular structures resulting from the freezing of water goes back over a century, but the first reported instance of freeze-casting, in the modern sense, was in 1954 when Maxwell et al. attempted to fabricate turbosupercharger blades out of refractory powders. They froze extremely thick slips of titanium carbide, producing near-net-shape castings that were easy to sinter and machine. The goal of this work, however, was to make dense ceramics. It was not until 2001, when Fukasawa et al. created directionally porous alumina castings, that the idea of using freeze-casting as a means of creating novel porous structures really took hold. Since that time, research has grown considerably with hundreds of papers of papers coming out within the last decade.

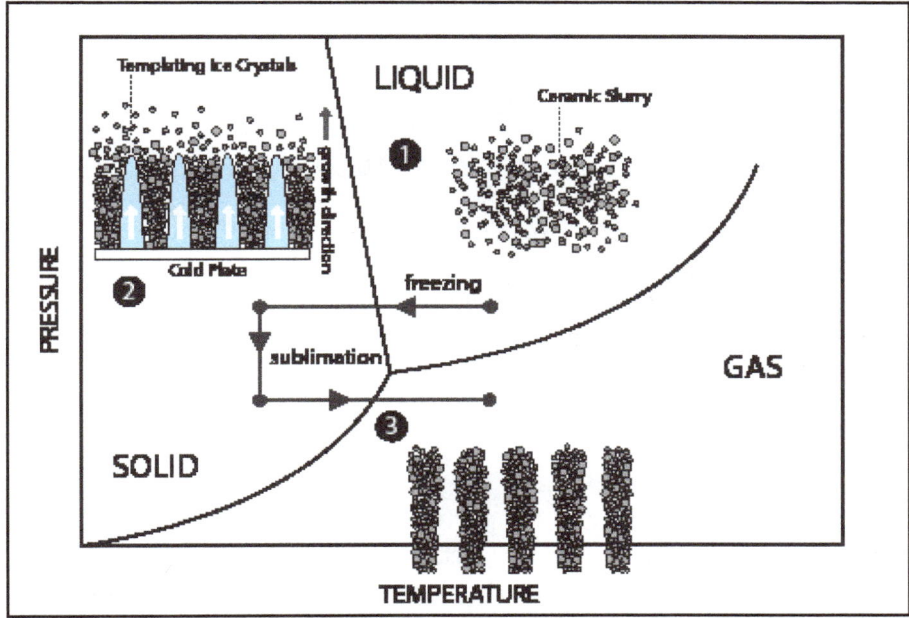

Steps in the freeze-casting process.

Because freeze-casting is a physical process, the techniques developed for one material system can be applied to a wide range of materials. Additionally, due to the inordinate amount of control and broad range of possible porous microstructures that freeze-casting can produce, the technique has found its niche in a number of disparate fields such as tissue scaffolds, photonics, metal-matrix composites, dentistry, materials science, and even food science.

There are three possible end results to uni-directionally freezing a suspension of particles. First, the ice-growth proceeds as a planar front, pushing particles in front like a bulldozer pushes a pile of rocks. This scenario usually occurs at very low solidification velocities (< 1 μm s⁻¹) or with extremely fine particles because they can move by Brownian motion away from the front. The resultant structure contains no macroporosity. If one were to increase the solidification speed, the size of the particles or solid loading moderately, the particles begin to interact in a meaningful way with the approaching ice front. The result is typically a lamellar or cellular templated structure whose exact morphology depends on the particular conditions of the system. It is this type of solidification that is targeted for porous materials made by freeze-casting. The third possibility for a freeze-cast structure occurs when particles are given insufficient time to segregate from the suspension, resulting in complete encapsulation of the particles within the ice front. This occurs when the freezing rates are rapid, particle size becomes sufficiently large, or when the solids loading is high enough to hinder particle motion. To ensure templating, the particles must be ejected from the oncoming front. Energetically speaking, this will occur if there is an overall increase in free energy if the particle were to be engulfed $(\Delta\sigma > 0)$.

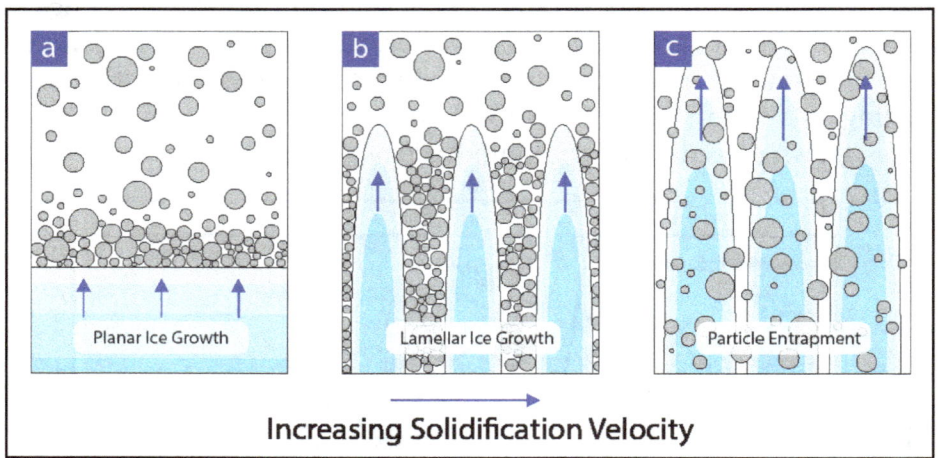

Depending on the speed of the freezing front, particle size and solids loading there are three pos-sible morphological outcomes: (a) planar front where all particles are pushed ahead of the ice, (b) lamellar/cellular front where ice crystals template particles or (c) particles are engulfed producing no ordering.

$$\Delta\sigma = \sigma_{ps} - (\sigma_{pl} + \sigma_{sl})$$

where $\Delta\sigma$ is the change in free energy of the particle, is the surface potential between the particle and interface, σ_{pl} is the potential between the particle and the liquid phase and σ_{sl} is the surface potential between the solid and liquid phases. This expression is valid at low solidification velocities, when the system is shifted only slightly from equilibrium. At high solidification velocities, kinetics must also be taken into consideration. There will be a liquid film between the front and particle to maintain constant transport of the molecules which are incorporated into the growing crystal. When the front velocity increases, this film thickness (d) will decrease due to increasing drag forces. A critical velocity (v_c) occurs when the film is no longer thick enough to supply the needed molecular supply. At this speed the particle will be engulfed. Most authors express v_c as a function of particle size where $v_c \propto \dfrac{1}{R}$. The transition from a porous R (lamellar) morphology to one where the majority of particles are entrapped occurs at v_c, which was defined by Deville et al. to be:

$$v_c = \frac{\Delta\sigma d}{3\eta R}\left(\frac{a_0}{d}\right)^z$$

where a_0 is the average intermolecular distance of the molecule that is freezing within the liquid, d is the overall thickness of the liquid film, η is the solution viscosity, R is the particle radius and

z is an exponent that can vary from 1 to 5. As expected, we see that v_c decreases as particle radius R goes up.

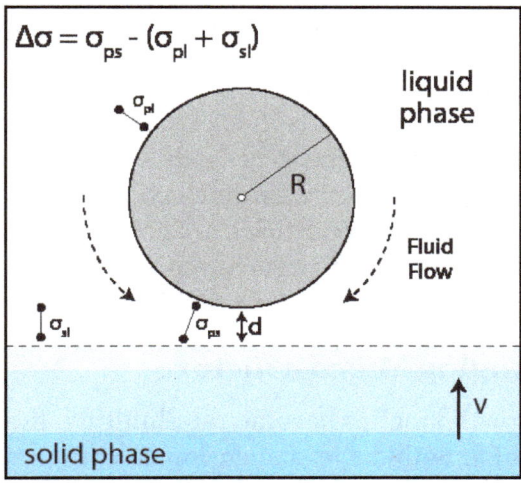

Schematic of a particle within the liquid phase interacting with an oncoming solidification front.

Waschkies et al. studied the structure of dilute to concentrated freeze-casts from low (< 1 μm s⁻¹) to extremely high (> 700 μm s⁻¹) solidification velocities. From this study, they were able to generate morphological maps for freeze-cast structures made under various conditions. Maps such as these are excellent for showing general trends, but they are quite specific to the materials system from which they were derived. For most applications where freeze-casts will be used after freezing, binders are needed to supply strength in the green state. The addition of binder can significantly alter the chemistry within the frozen environment, depressing the freezing point and hampering particle motion leading to particle entrapment at speeds far below the predicted v_c. Assuming, however, that we are operating at speeds below VC and above those which produce a planar front, we will achieve some cellular structure with both ice-crystals and walls composed of packed ceramic particles. The morphology of this structure is tied to some variables, but the most influential is the temperature gradient as a function of time and distance along the freezing direction.

Freeze-casts have at least three apparent morphological regions. At the side where freezing initiates is a nearly isotropic region with no visible macropores dubbed the Initial Zone (IZ). Directly after the IZ is the Transition Zone (TZ), where macropores begin to form and align with one another. The pores in this region may appear randomly oriented. The third zone is called the Steady-State Zone (SSZ), macropores in this region are aligned with one another and grow in a regular fashion. Within the SSZ, the structure is defined by a value λ that is the average thickness of a ceramic wall and its adjacent macropore.

Initial Zone: Nucleation and Growth Mechanisms

Although the ability of ice to exclude suspended particles has long been known, the mechanism is still being debated. It was believed initially that during the moments immediately following the nucleation of the ice crystals, particles were ejected from the growing planar ice front, leading to the formation of a constitutionally super-cooled zone directly ahead of the growing ice. This unstable region eventually resulted in perturbations, breaking the planar front into a columnar ice front, a phenomenon better known as a Mullins-Serkerka instability. After the breakdown, the ice crystals grow along the temperature gradient, pushing ceramic particles from the liquid phase aside so that they accumulate between the growing ice crystals. However, recent in-situ X-ray radiography of directionally frozen alumina suspensions reveal a different mechanism.

In-situ testing reveals that freeze-casting is an aggressive growth process. In the moments immediately before nucleation, the suspension is in an unstable super-cooled state. Homogeneous (spatially speaking) nucleation of ice crystals occurs followed by explosive crystal growth in every spatial and crystallographic direction. The initial nucleation and growth steps are so rapid (approaching 800 mm s^{-1}) that all suspended particles are completely engulfed by the oncoming ice front because not enough time is given for particle redistribution, resulting in a structure with anisotropic particle distribution. This step is what provides the initial zone structure.

Transition Zone: A Changing Microstructure

As solidification slows and growth kinetics become rate-limiting, the ice crystals begin to exclude the particles, redistributing them within the suspension. A competitive growth process develops between two crystal populations, those with their basal planes aligned with the thermal gradient (z-crystals) and those that are randomly oriented (r-crystals) giving rise to the start of the TZ.

There are colonies of similarly aligned ice crystals growing throughout the suspension. There are fine lamellae of aligned z-crystals growing with their basal planes aligned with the thermal gradient. The r-crystals appear in this cross-section as platelets but in actuality, they are most similar to columnar dendritic crystals cut along a bias. Within the transition zone, the r-crystals either stop growing or turn into z-crystals that eventually become the predominant orientation, and lead to steady-state growth. There are some reasons why this occurs. For one, during freezing, the growing crystals tend to align with the temperature gradient, as this is the lowest energy configuration and thermodynamically preferential. Aligned growth, however, can mean two different things. Assuming the temperature gradient is vertical, the growing crystal will either be parallel (z-crystal) or perpendicular (r-crystal) to this gradient. A crystal that lays horizontally can still grow in line with the temperature gradient, but it will mean growing on its face rather than its edge. Since the thermal conductivity of ice is so small (1.6 - 2.4 W mK^{-1}) compared with most every other ceramic (ex. Al$_2$O$_3$ = 40 W mK^{-1}), the growing ice will have a significant insulative effect on the localized thermal conditions within the slurry. This can be illustrated using simple resistor elements.

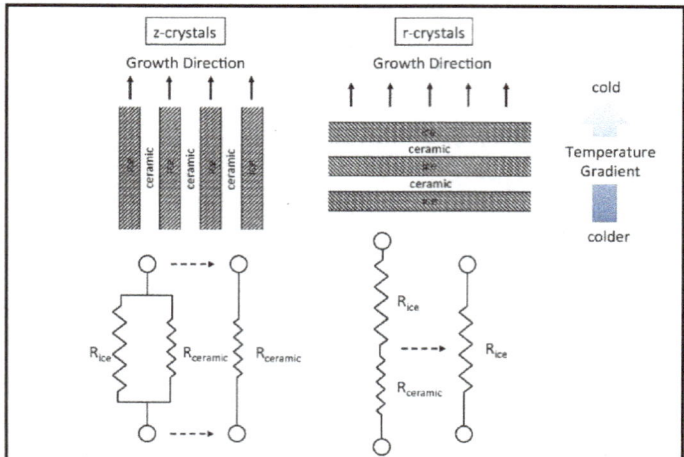

Shows thermal resistance of the two extreme cases of crystallographic alignment.

When ice crystals are aligned with their basal planes parallel to the temperature gradient (z-crystals), they can be represented as two resistors in parallel. The thermal resistance of the ceramic is significantly smaller than that of the ice however, so the apparent resistance can be expressed as the lower $R_{ceramic}$. If the ice crystals are aligned perpendicular to the temperature gradient (r-crystals), they can be approximated as two resistor elements in series. For this case, the R_{ice} is limiting and will dictate the localized thermal conditions. The lower thermal resistance for the z-crystal case leads to lower temperatures and greater heat flux at the growing crystals tips, driving further growth in this direction while, at the same time, the large R_{ice} value hinders the growth of the r-crystals. Each ice crystal growing within the slurry will be some combination of these two scenarios. Thermodynamics dictate that all crystals will tend to align with the preferential temperature gradient causing r-crystals to eventually give way to z-crystals, which can be seen from the following radiographs taken within the TZ.

When z-crystals become the only significant crystal orientation present, the ice-front grows in a steady-state manner except there are no significant changes to the system conditions. It was observed in 2012 that, in the initial moments of freezing, there are dendritic r-crystals that grow 5 - 15 times faster than the solidifying front. These shoot up into the suspension ahead of the main ice front and partially melt back. Interestingly, these crystals stop growing at the point where the TZ will eventually fully transition to the SSZ. Researchers determined that this particular point marks the position where the suspension is in an equilibrium state (i.e. freezing temperature and suspension temperature are equal). We can say then that the size of the initial and transition zones are controlled by the extent of supercooling beyond the already low freezing temperature. If the freeze-casting setup is controlled so that nucleation is favored at only small supercooling, then the TZ will give way to the SSZ sooner.

Steady-state Growth Zone

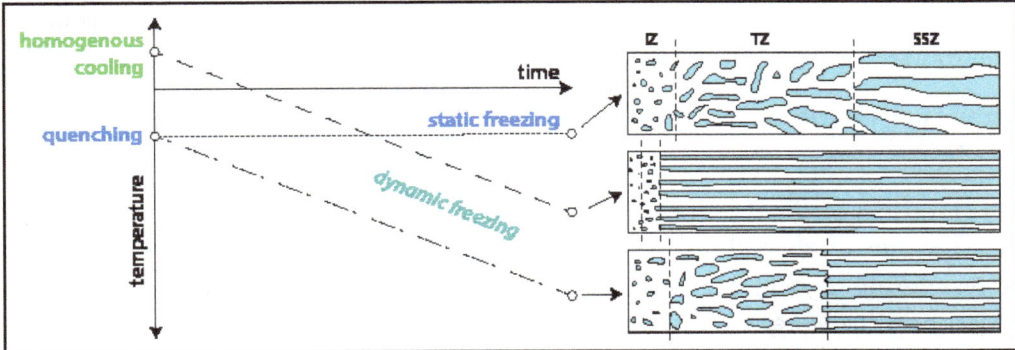

Shows various thermal profiles and their effect on subsequent microstructure of freeze-casts.

The structure in this final region contains long, aligned lamellae that alternate between ice crystals and ceramic walls. The faster a sample is frozen, the finer its solvent crystals (and its eventual macroporosity) will be. Within the SSZ, the normal speeds which are usable for colloidal templating are $10 - 100$ mm s^{-1} leading to solvent crystals typically between 2 mm and 200 mm. Subsequent sublimation of the ice within the SSZ yields a green ceramic preform with porosity in a nearly exact replica of these ice crystals. The microstructure of a freeze-cast within the SSZ is defined by its wavelength (λ) which is the average thickness of a single ceramic wall plus its adjacent macropore. Several publications have reported the effects of solidification kinetics on the microstructures of freeze-cast materials. It has been shown that λ follows an empirical power-law relationship with solidification velocity (v) (Eq. 2.14):

$$\lambda = Av^{-n}$$

Both A and v are used as fitting parameters as currently there is no way of calculating them from first principles, although it is generally believed that A is related to slurry parameters like viscosity and solid loading while n is influenced by particle characteristics.

Controlling the Porous Structure

Stop-motion animation of the freeze-casting process.

There are two general categories of tools for architecture a freeze-cast:

1. Chemistry of the System - freezing medium and chosen particulate material(s), any additional binders, dispersants or additives.

2. Operational Conditions - temperature profile, atmosphere, mold material, freezing surface, etc.

Initially, the materials system is chosen based on what sort of final structure is needed. This review has focused on water as the vehicle for freezing, but there are some other solvents that may be used. Notably, camphene, which is an organic solvent that is waxy at room temperature. Freezing of this solution produces highly branched dendritic crystals. Once the materials system is settled on however, the majority of microstructural control comes from external operational conditions such as mold material and temperature gradient.

Controlling Pore Size

The microstructural wavelength (average pore + wall thickness) can be described as a function of the solidification velocity v ($l = Av^{-n}$) where A is dependent on solids loading. There are two ways then that the pore size can be controlled. The first is to change the solidification speed that then alters the microstructural wavelength, or the solids loading can be changed. In doing so, the ratio of pore size to wall size is changed. It is often more prudent to alter the solidification velocity seeing as a minimum solid loading is usually desired. Since microstructural size (λ) is inversely related to the velocity of the freezing front, faster speeds lead to finer structures, while slower speeds produce a coarse microstructure. Controlling the solidification velocity is, therefore, crucial to being able to control the microstructure.

Controlling Pore Shape

Additives can prove highly useful and versatile in changing the morphology of pores. These work by affecting the growth kinetics and microstructure of the ice in addition to the topology of the ice-water interface. Some additives work by altering the phase diagram of the solvent. For example, water and NaCl have a eutectic phase diagram. When NaCl is added into a freeze-casting suspension, the solid ice phase and liquid regions are separated by a zone where both solids and liquids can coexist. This briny region is removed during sublimation, but its existence has a strong effect on the microstructure of the porous ceramic. Other additives work by either altering the interfacial surface energies between the solid/liquid and particle/liquid, changing the viscosity of the suspension, or the degree of undercooling in the system. Studies have been done with glycerol, sucrose, ethanol, coca-cola, acetic acid and more.

Static vs. Dynamic Freezing Profiles

If a freeze casting setup with a constant temperature on either side of the freezing system is used, (static freeze-casting) the front solidification velocity in the SSZ will decrease over time due to the increasing thermal buffer caused by the growing ice front. When this occurs, more time is given for the anisotropic ice crystals to grow perpendicularly to the freezing direction (c-axis) resulting in a structure with ice lamellae that increase in thickness along the length of the sample

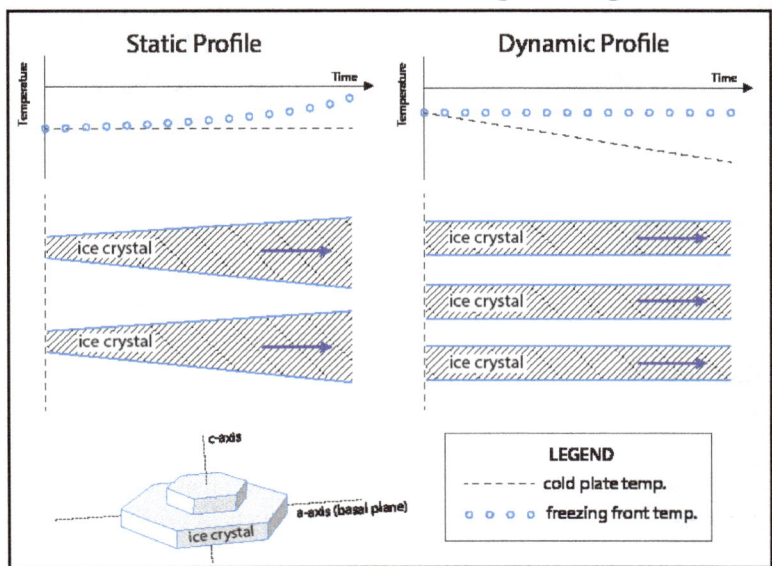

Static and dynamic freezing profiles in the steady-state freezing regime.

To ensure highly anisotropic, yet predictable solidification behavior within the SSZ, dynamic freezing patterns are preferred. Using dynamic freezing, the velocity of the solidification front, and, therefore, the ice crystal size, can be controlled with a changing temperature gradient. The increasing thermal gradient counters the effect of the growing thermal buffer imposed by the growing ice front. It has been shown that a linearly decreasing temperature on one side of a freeze-cast will result in near-constant solidification velocity, yielding ice crystals with an almost constant thickness along the SSZ of an entire sample. However, as pointed out by Waschkies et al. even with constant solidification velocity, the thickness of the ice crystals does increase slightly over the course of freezing.

Anisotropy of the Interface Kinetics

Even if the temperature gradient within the slurry is perfectly vertical, it is common to see tilting or curvature of the lamellae as they grow through the suspension. To explain this, it is possible to define two distinct growth directions for each ice crystal. There is the direction determined by the temperature gradient, and the one defined by the preferred growth direction crystallographically speaking. These angles are often at odds with one another, and their balance will describe the tilt of the crystal.

The non-overlapping growth directions also help to explain why dendritic textures are often seen in freeze-casts. This texturing is usually found only on the side of each lamella; the direction of the imposed temperature gradient. The ceramic structure left behind shows the negative image of these dendrites. In 2013, Deville et al. made the interesting observation that the periodicity of these dendrites (tip-to-tip distance) actually seems to be related to the primary crystal thickness.

Particle Packing Effects

Up until now, the focus has been mostly on the structure of the ice itself; the particles are almost an afterthought to the templating process but in fact, the particles can and do play a significant role during freeze-casting. It turns out that particle arrangement also changes as a function of the freezing conditions. For example, researchers have shown that freezing velocity has a marked effect on wall roughness. Faster freezing rates produce rougher walls since particles are given insufficient time to rearrange. This could be of use when developing permeable gas transfer membranes where tortuosity and roughness could impede gas flow. It also turns out that z- and r-crystals do not interact with ceramic particles in the same way. The z-crystals pack particles in the x-y plane while r-crystals pack particles primarily in the z-direction. R-crystals actually pack particles more efficiently than z-crystals and because of this, the area fraction of the particle-rich phase (1 - area fraction of ice crystals) changes as the crystal population shifts from a mixture of z- and r-crystals to only z-crystals. Starting from where ice crystals first begin to exclude particles, marking the beginning of the transition zone, we have a majority of r-crystals and a high value for the particle-rich phase fraction. We can assume that because the solidification speed is still rapid that the particles will not be packed efficiently. As the solidification rate slows down, however, the area fraction of the particle-rich phase drops indicating an increase in packing efficiency. At the same time, the competitive growth process is taking place, replacing r-crystals with z-crystals. At a certain point nearing the end of the transition zone, the particle-rich phase fraction rises sharply since z-crystals are less efficient at packing particles than r-crystals. The apex of this curve marks the point where only z-crystals are present (SSZ). During steady-state growth, after the maximum particle-rich phase fraction is reached, the efficiency of packing increases as steady-state is achieved. In 2011, researchers at Yale University set out to probe the actual spatial packing of particles within the walls. Using small-angle X-ray scattering (SAXS) they characterized the particle size, shape and interparticle spacing of nominally 32 nm silica suspensions that had been freeze-cast at different speeds. Computer simulations indicated that for this system, the particles within the walls should not be touching but rather separated from one another by thin films of ice. Testing however revealed that the particles were, in fact, touching and more than that, they attained a packed morphology that cannot be explained by typical equilibrium densification processes.

Morphological Instabilities

In an ideal world, the spatial concentration of particles within the SSZ would remain constant throughout solidification. As it happens, though, the concentration of particles does change during compression, and this process is highly sensitive to solidification speed. At low freezing rates, Brownian motion takes place, allowing particles to move easily away from the solid-liquid interface and maintain a homogeneous suspension. In this situation, the suspension is always warmer than the solidified portion. At fast solidification speeds, approaching VC, the concentration, and concentration gradient at the solid-liquid interface increases because particles cannot redistribute soon enough. When it has built up enough, the freezing point of the suspension is below the temperature gradient in the solution and morphological instabilities can occur. For situations where the particle concentration bleeds into the diffusion layer, both the actual and freezing temperature dip below the equilibrium freezing temperature creating an unstable system. Often, these situations lead to the formation of what are known as ice lenses.

These morphological instabilities can trap particles, preventing full redistribution and resulting in inhomogeneous distribution of solids along the freezing direction as well as discontinuities in the ceramic walls, creating voids larger than intrinsic pores within the walls of the porous ceramic.

Novel Freeze-Casting Techniques

Freeze-casting can be applied to numerous materials systems including ceramics, polymers and metals. As long as there are particles that may be excluded when the solvent changes phase, a templated structure is possible. Using various novel processing techniques, some authors have demonstrated even greater levels of control made available with freeze-casting. Munch et al. showed that it is possible to control the long-range arrangement and orientation of crystals normal to the growth direction by templating the nucleation surface. This technique works by providing lower energy nucleation sites to control the initial crystal growth and arrangement. The orientation of ice crystals can also be affected by applied electromagnetic fields as was demonstrated in 2010 by Tang et al. Using specialized setups, researchers have been able to create radially aligned freeze-casts tailored for filtration or gas separation applications. Inspired by Nature, scientists have also been able to use coordinating chemicals and cryopreserved to create remarkably distinctive microstructural architectures.

Freeze-gelation

Freeze-gelation, is a form of sol-gel processing of ceramics that enables a ceramic object to be fabricated in complex shapes, without the need for high-temperature sintering. The process is similar to freeze-casting.

The process is simple, but the science is, as of 2005, not well understood. The most common process involves the mixing of a silica solution with a filler powder. For example, if we were making a component out of alumina, aluminium oxide, then we would still use a silica sol, but alumina filler powder. The relative amounts used differ, normally between 3 and 4 times more filler than sol is added by weight.

A wetting agent is added, such that the filler powder disperses properly in the sol, which is mostly water. This makes the mixture doughy and stiff. The mixture is, however, highly thixotropic, so that when vibrated it turns liquid. The stiff dough is placed in a mold and the mold vibrated to liquefy the mixture, filling the mold and releasing any trapped air.

The filled mold is then frozen. On freezing, silica precipitates from the sol, forming a gel. This gel holds the filler powder together in something approximating a sintering greenform. The component is then dried in a furnace, leaving the component.

The advantages of freeze-geleation over sintering are essentially cost-based. It doesn't require high pressure equipment or powerful furnaces (drying temperatures are only just above water's boiling point), yet it creates a useful product which takes the shape of the mold very accurately.

History

In terms of being simply a process by which powder can be made into a monolith, freeze casting could be as old as the earth. A material called laminar opaline silica or LOS is believed to be formed by the freeze casting of volcanic ash, some soils containing the required sols to make the gel.

Artificially it is also an old process, having been known and studied for 100 years or more, but never brought to significant industrial application. Lottermoser, a German, wrote a paper on 'das Ausfrieren von Hydrosolen' (the Freezing of Hydrosols) in 1908. Through the 20th century various people have patented techniques using freeze-gelation, most being centred on the use of ceramics as refractory materials. A furnace lining brick, or an investment casting mold, can be easily fabricated using this method.

Recently there was a flurry of interest in freeze-casting at the University of Bath, UK, which led most significantly to two doctoral theses, by J. Laurie in 1995 and by M. Statham in 1998. Taken together in chronological order, these form a good introduction to the technique for the interested party.

Applications

To consider the applications of freeze-casting, we should consider the properties of the freeze-cast component. First, and critically, it is not fully dense. It contains only about 60–70% solid matter, the remainder being air in the form of porosity. This is turn leads to an interesting property of freeze-castings – they are often porous, not merely at the surface, but throughout their thickness. A fluid will penetrate through the pores in the casting and eventually soak through, like a sponge. This is because at porosity percentages above the 'pore percolation threshold', pores link up into continuous channels. The pore percolation threshold depends on the characteristics of the material, but it is normally very roughly around 20%. A 60% dense component has 40% porosity.

As we might expect, this amount of air in the component reduces its strength a lot. Pure, fully dense alumina, for example, is as strong as steel – far stronger if processed carefully – but freeze-cast alumina components are of similar strength to concrete. The freeze-cast component also tends to be brittle, fracturing easily.

It is unlikely then that freeze-cast components could be used structurally (without further processing – more later), but they have other properties that make them useful. They are rather light, with freeze-cast alumina components having a density somewhere in the region of 2.5 g/cm^3, similar to aluminium. They are easy and cheap to make, from inexpensive and safe ingredients and using no dangerous equipment. They can take complex shapes, as they are cast, rather than machined. They can also be very large, probably larger than monolithic ceramic components made by any other process. Finally, and crucially, their porosity means that they can be infiltrated by materials with useful properties, or processed with other materials in. For example, the component could be dipped in molten copper, such that the copper is drawn up by capillary action into the porosity, increasing the conductivity of the component vastly. Alternatively, copper powder could be used as a filler powder in place of some alumina to the same end.

Freeze-cast components, in their basic form, are ideal for use as heat-resisting objects. In this way, they can be useful in metalwork, as molds or as substrates for metal spray-forming. However, with suitable post-processing, they could fulfil many other applications, such as silicon chip mounts, or even engine blocks.

Theory

The science is not particularly well understood. It has been known for years that silica sols (also known as colloidal silica, silicic acid, polysilicic acid) will gel when exposed to temperatures around 0 °C (32 °F). The theoretical mechanism is quite simple:

Colloidal silica is produced by the polymerisation of monosilicic acid, $Si(OH)_4$, until the chains of polysilicic acid become so long they form silica particles with hydroxylated surfaces. On freezing of the sol, the silica particles are rejected away from the solidifying interface and forced into the interstices between the ice crystals. Here, they come into contact with each other, and link via the condensation of their surface hydroxyl groups into siloxane bonds. This, happening throughout the sol, forms a gel.

In a filled sol, the ceramic powder is trapped within the gel, and forms a monolith.

Sintering

Sintering is the process of compacting and forming a solid mass of material by heat or pressure without melting it to the point of liquefaction.

Sintering happens naturally in mineral deposits or as a manufacturing process used with metals, ceramics, plastics, and other materials. The atoms in the materials diffuse across the boundaries of the particles, fusing the particles together and creating one solid piece. Because the sintering temperature does not have to reach the melting point of the material, sintering is often chosen as the shaping process for materials with extremely high melting points such as tungsten and molybdenum. The study of sintering in metallurgy powder-related processes is known as powder metallurgy. An example of sintering can be observed when ice cubes in a glass of water adhere to each other, which is driven by the temperature difference between the water and the ice. Examples of pressure-driven sintering are the compacting of snowfall to a glacier, or the forming of a hard snowball by pressing loose snow together.

The word "sinter" comes from the Middle High German *sinter*, a cognate of English "cinder".

Clinker nodules produced by sintering.

General Sintering

Sintering is effective when the process reduces the porosity and enhances properties such as strength, electrical conductivity, translucency and thermal conductivity; yet, in other cases, it may be useful to increase its strength but keep its gas absorbency constant as in filters or catalysts. During the firing process, atomic diffusion drives powder surface elimination in different stages, starting from the formation of necks between powders to final elimination of small pores at the end of the process.

The driving force for densification is the change in free energy from the decrease in surface area and lowering of the surface free energy by the replacement of solid-vapor interfaces. It forms new but lower-energy solid-solid interfaces with a total decrease in free energy occurring on sintering 1-micrometre particles a 1 cal/g decrease. On a microscopic scale, material transfer is affected by the change in pressure and differences in free energy across the curved surface. If the size of the particle is small (and its curvature is high), these effects become very large in magnitude. The change in energy is much higher when the radius of curvature is less than a few micrometres, which is one of the main reasons why much ceramic technology is based on the use of fine-particle materials.

For properties such as strength and conductivity, the bond area in relation to the particle size is the determining factor. The variables that can be controlled for any given material are the temperature and the initial grain size, because the vapor pressure depends upon temperature. Through time, the particle radius and the vapor pressure are proportional to $(p_o)^{2/3}$ and to $(p_o)^{1/3}$, respectively.

The source of power for solid-state processes is the change in free or chemical potential energy between the neck and the surface of the particle. This energy creates a transfer of material through the fastest means possible; if transfer were to take place from the particle volume or the grain boundary between particles, then there would be particle reduction and pore destruction. The pore elimination occurs faster for a trial with many pores of uniform size and higher porosity where the boundary diffusion distance is smaller. For the latter portions of the process, boundary and lattice diffusion from the boundary become important.

Control of temperature is very important to the sintering process, since grain-boundary diffusion and volume diffusion rely heavily upon temperature, the size and distribution of particles of the material, the materials composition, and often the sintering environment to be controlled.

Ceramic Sintering

Sintering is part of the firing process used in the manufacture of pottery and other ceramic objects. These objects are made from substances such as glass, alumina, zirconia, silica, magnesia, lime, beryllium oxide, and ferric oxide. Some ceramic raw materials have a lower affinity for water and a lower plasticity index than clay, requiring organic additives in the stages before sintering. The general procedure of creating ceramic objects via sintering of powders includes:

- Mixing water, binder, deflocculant, and unfired ceramic powder to form a slurry;

- Spray-drying the slurry;

- Putting the spray dried powder into a mold and pressing it to form a green body (an unsintered ceramic item);

- Heating the green body at low temperature to burn off the binder;

- Sintering at a high temperature to fuse the ceramic particles together.

All the characteristic temperatures associated with phase transformation, glass transitions, and melting points, occurring during a sinterisation cycle of a particular ceramics formulation (i.e., tails and frits) can be easily obtained by observing the expansion-temperature curves during optical dilatometer thermal analysis. In fact, sinterisation is associated with a remarkable shrinkage of the material because glass phases flow once their transition temperature is reached, and start consolidating the powdery structure and considerably reducing the porosity of the material.

Sintering is performed at high temperature. Besides, second and/or third external force (such as pressure, electrical current) could be used. Commonly used second external force is pressure. So, the sintering that performed just using temperature is generally called "pressureless sintering". Pressureless sintering is possible with graded metal-ceramic composites, with a nanoparticle sintering aid and bulk molding technology. A variant used for 3D shapes is called hot isostatic pressing.

To allow efficient stacking of product in the furnace during sintering and prevent parts sticking together, many manufacturers separate ware using ceramic powder separator sheets. These sheets are available in various materials such as alumina, zirconia and magnesia. They are additionally categorized by fine, medium and coarse particle sizes. By matching the material and particle size to the ware being sintered, surface damage and contamination can be reduced while maximizing furnace loading.

Sintering of Metallic Powders

Iron powder.

Most, if not all, metals can be sintered. This applies especially to pure metals produced in vacuum which suffer no surface contamination. Sintering under atmospheric pressure requires the use of a protective gas, quite often endothermic gas. Sintering, with subsequent reworking, can produce a great range of material properties. Changes in density, alloying, or heat treatments can alter the physical characteristics of various products. For instance, the Young's Modulus E_n of sintered iron powders remains insensitive to sintering time, alloying, or particle size in the original powder, but depends upon the density of the final product:

$$E_n / E = (D / d)^{3.4},$$

where D is the density, E is Young's modulus and d is the maximum density of iron.

Sintering is static when a metal powder under certain external conditions may exhibit coalescence, and yet reverts to its normal behavior when such conditions are removed. In most cases, the density of a collection of grains increases as material flows into voids, causing a decrease in overall volume. Mass movements that occur during sintering consist of the reduction of total porosity by repacking, followed by material transport due to evaporation and condensation from diffusion. In the final stages, metal atoms move along crystal boundaries to the walls of internal pores, redistributing mass from the internal bulk of the object and smoothing pore walls. Surface tension is the driving force for this movement.

A special form of sintering (which is still considered part of powder metallurgy) is liquid-state sintering in which at least one but not all elements are in a liquid state. Liquid-state sintering is required for making cemented carbide or tungsten carbide.

Sintered bronze in particular is frequently used as a material for bearings, since its porosity allows lubricants to flow through it or remain captured within it. Sintered copper may be used as a wicking structure in certain types of heat pipe construction, where the porosity allows a liquid agent to move through the porous material via capillary action. For materials that have high melting points such as molybdenum, tungsten, rhenium, tantalum, osmium and carbon, sintering is one of the few viable manufacturing processes. In these cases, very low porosity is desirable and can often be achieved.

Sintered metal powder is used to make frangible shotgun shells called breaching rounds, as used by military and SWAT teams to quickly force entry into a locked room. These shotgun shells are designed to destroy door deadbolts, locks and hinges without risking lives by ricocheting or by flying on at lethal speed through the door. They work by destroying the object they hit and then dispersing into a relatively harmless powder.

Sintered bronze and stainless steel are used as filter materials in applications requiring high temperature resistance while retaining the ability to regenerate the filter element. For example, sintered stainless steel elements are employed for filtering steam in food and pharmaceutical applications, and sintered bronze in aircraft hydraulic systems.

Sintering of powders containing precious metals such as silver and gold is used to make small jewelry items.

Advantages

Particular advantages of the powder technology include:

1. Very high levels of purity and uniformity in starting materials

2. Preservation of purity, due to the simpler subsequent fabrication process (fewer steps) that it makes possible

3. Stabilization of the details of repetitive operations, by control of grain size during the input stages

4. Absence of binding contact between segregated powder particles – or "inclusions" (called stringering) – as often occurs in melting processes

5. No deformation needed to produce directional elongation of grains

6. Capability to produce materials of controlled, uniform porosity

7. Capability to produce nearly net-shaped objects

8. Capability to produce materials which cannot be produced by any other technology

9. Capability to fabricate high-strength material like turbine blades

10. After sintering the mechanical strength to handling becomes higher

The literature contains many references on sintering dissimilar materials to produce solid/solid-phase compounds or solid/melt mixtures at the processing stage. Almost any substance can be obtained in powder form, through either chemical, mechanical or physical processes, so basically any material can be obtained through sintering. When pure elements are sintered, the leftover powder is still pure, so it can be recycled.

Disadvantages

Particular disadvantages of the powder technology include:

1. 100% sintered (iron ore) can not be charged in the blast furnace.

2. By sintering one cannot create uniform sizes.

3. Micro- and nano-structures produced before sintering are often destroyed.

Plastics Sintering

Plastic materials are formed by sintering for applications that require materials of specific porosity. Sintered plastic porous components are used in filtration and to control fluid and gas flows. Sintered plastics are used in applications requiring wicking properties, such as marking pen nibs. Sintered ultra high molecular weight polyethylene materials are used as ski and snowboard base materials. The porous texture allows wax to be retained within the structure of the base material, thus providing a more durable wax coating.

Liquid Phase Sintering

For materials which are difficult to sinter, a process called liquid phase sintering is commonly used. Materials for which liquid phase sintering is common are Si_3N_4, WC, SiC, and more. Liquid phase sintering is the process of adding an additive to the powder which will melt before the matrix phase. The process of liquid phase sintering has three stages:

- Rearrangement – As the liquid melts capillary action will pull the liquid into pores and also cause grains to rearrange into a more favorable packing arrangement.

- Solution-Precipitation – In areas where capillary pressures are high (particles are close together) atoms will preferentially go into solution and then precipitate in areas of lower chemical potential where particles are not close or in contact. This is called "contact flattening". This densifies the system in a way similar to grain boundary diffusion in solid state sintering. Ostwald ripening will also occur where smaller particles will go into solution preferentially and precipitate on larger particles leading to densification.

- Final Densification – densification of solid skeletal network, liquid movement from efficiently packed regions into pores.

For liquid phase sintering to be practical the major phase should be at least slightly soluble in the liquid phase and the additive should melt before any major sintering of the solid particulate network occurs, otherwise rearrangement of grains will not occur. Liquid phase sintering was successfully applied to improve grain growth of thin semiconductor layers from nanoparticle precursor films.

Electric Current Assisted Sintering

These techniques employ electric currents to drive or enhance sintering. English engineer A. G. Bloxam registered in 1906 the first patent on sintering powders using direct current in vacuum. The primary purpose of his inventions was the industrial scale production of filaments for incandescent lamps by compacting tungsten or molybdenum particles. The applied current was particularly effective in reducing surface oxides that increased the emissivity of the filaments.

In 1913, Weintraub and Rush patented a modified sintering method which combined electric current with pressure. The benefits of this method were proved for the sintering of refractory metals as well as conductive carbide or nitride powders. The starting boron–carbon or silicon–carbon powders were placed in an electrically insulating tube and compressed by two rods which also served as electrodes for the current. The estimated sintering temperature was 2000 °C.

In the United States, sintering was first patented by Duval d'Adrian in 1922. His three-step process aimed at producing heat-resistant blocks from such oxide materials as zirconia, thoria or tantalia. The steps were: (i) molding the powder; (ii) annealing it at about 2500 °C to make it conducting; (iii) applying current-pressure sintering as in the method by Weintraub and Rush.

Sintering that uses an arc produced via a capacitance discharge to eliminate oxides before direct current heating, was patented by G. F. Taylor in 1932. This originated sintering methods employing pulsed or alternating current, eventually superimposed to a direct current. Those techniques have been developed over many decades and summarized in more than 640 patents.

Of these technologies the most well known is resistance sintering (also called hot pressing) and spark plasma sintering, while Electro Sinter Forging is the latest advancement in this field.

Spark Plasma Sintering

In spark plasma sintering (SPS), external pressure and an electric field are applied simultaneously to enhance the densification of the metallic/ceramic powder compacts. This densification uses lower temperatures and shorter amount of time than typical sintering. For a number of years, it was speculated that the existence of sparks or plasma between particles could aid sintering; however, Hulbert and coworkers systematically proved that the electric parameters used during spark plasma sintering make it (highly) unlikely. In light of this, the name "spark plasma sintering" has been rendered obsolete. Terms such as "Field Assisted Sintering Technique" (FAST), "Electric Field Assisted Sintering" (EFAS), and Direct Current Sintering (DCS) have been implemented by the sintering community. Using a DC pulse as the electric current, spark plasma, spark impact pressure, joule heating, and an electrical field diffusion effect would be created.

Electro Sinter Forging

Electro Sinter Forging is an electric current assisted sintering (ecas) technology originated from Capacitor discharge sintering. It is used for the production of diamond metal matrix composites and under evaluation for the production of hard metals, nitinol and other metals and intermetallics. It is characterized by a very low sintering time allowing machines to sinter at the same speed as a compaction press.

Pressureless Sintering

Pressureless sintering is the sintering of a powder compact (sometimes at very high temperatures, depending on the powder) without applied pressure. This avoids density variations in the final component, which occurs with more traditional hot pressing methods.

The powder compact (if a ceramic) can be created by slip casting, injection moulding, and cold isostatic pressing. After pre-sintering, the final green compact can be machined to its final shape before sintered.

Three different heating schedules can be performed with pressureless sintering: constant-rate of heating (CRH), rate-controlled sintering (RCS), and two-step sintering (TSS). The microstructure and grain size of the ceramics may vary depending on the material and method used.

Constant-rate of heating (CRH), also known as temperature-controlled sintering, consists of heating the green compact at a constant rate up to the sintering temperature. Experiments with zirconia have been performed to optimize the sintering temperature and sintering rate for CRH method. Results showed that the grain sizes were identical when the samples were sintered to the same density, proving that grain size is a function of specimen density rather than CRH temperature mode.

In rate-controlled sintering (RCS), the densification rate in the open-porosity phase is lower than in the CRH method. By definition, the relative density, ρ_{rel}, in open-porosity phase is lower than 90%. Although this should prevent separation of pores from grain boundaries, it has been proven statistically that RCS did not produce smaller grain sizes than CRH for alumina, zirconia, and ceria samples.

Two-step sintering (TSS) uses two different sintering temperatures. The first sintering temperature should guarantee a relative density higher than 75% of theoretical sample density. This will remove supercritical pores from the body. The sample will then be cooled down and held at the second sintering temperature until densification is completed. Grains of cubic zirconia and cubic strontium titanate were significantly refined by TSS compared to CRH. However, the grain size changes in other ceramic materials, like tetragonal zirconia and hexagonal alumina, were not statistically significant.

Microwave Sintering

In microwave sintering, heat is generated internally within the material, rather than via radiative heat transfer from an external heat source. Other benefits of microwave sintering are a better heat diffusion, less time needed to reach the sintering temperature, less heating energy required and improvements in the product properties.

As microwaves can only penetrate a short distance in materials with a high conductivity and a high permeability, microwave sintering requires the sample to be delivered in powders with a particle size around the penetration depth of microwaves in the particular material. The sintering process and side-reactions run several times faster during microwave sintering at the same temperature, which results in different properties for the sintered product.

This technique is acknowledged to be quite effective in maintaining fine grains/nano sized grains in sintered bioceramics. Magnesium phosphates and calcium phosphates are the examples which have been processed through microwave sintering technique.

Densification, Vitrification and Grain Growth

Sintering in practice is the control of both densification and grain growth. Densification is the act of reducing porosity in a sample thereby making it more dense. Grain growth is the process of grain boundary motion and Ostwald ripening to increase the average grain size. Many properties (mechanical strength, electrical breakdown strength, etc.) benefit from both a high relative density and a small grain size. Therefore, being able to control these properties during processing is of high technical importance. Since densification of powders requires high temperatures, grain growth naturally occurs during sintering. Reduction of this process is key for many engineering ceramics.

For densification to occur at a quick pace it is essential to have (1) an amount of liquid phase that is large in size, (2) a near complete solubility of the solid in the liquid, and (3) wetting of the solid by the liquid. The power behind the densification is derived from the capillary pressure of the liquid phase located between the fine solid particles. When the liquid phase wets the solid particles, each space between the particles becomes a capillary in which a substantial capillary pressure is developed. For submicrometre particle sizes, capillaries with diameters in the range of 0.1 to 1 micrometres develop pressures in the range of 175 pounds per square inch (1,210 kPa) to 1,750 pounds per square inch (12,100 kPa) for silicate liquids and in the range of 975 pounds per square inch (6,720 kPa) to 9,750 pounds per square inch (67,200 kPa) for a metal such as liquid cobalt.

Densification requires constant capillary pressure where just solution-precipitation material transfer would not produce densification. For further densification, additional particle movement while the particle undergoes grain-growth and grain-shape changes occurs. Shrinkage would result when the liquid slips between particles and increase pressure at points of contact causing the material to move away from the contact areas forcing particle centers to draw near each other.

The sintering of liquid-phase materials involves a fine-grained solid phase to create the needed capillary pressures proportional to its diameter and the liquid concentration must also create the required capillary pressure within range, else the process ceases. The vitrification rate is dependent upon the pore size, the viscosity and amount of liquid phase present leading to the viscosity of the overall composition, and the surface tension. Temperature dependence for densification controls the process because at higher temperatures viscosity decreases and increases liquid content. Therefore, when changes to the composition and processing are made, it will affect the vitrification process.

Sintering Mechanisms

Sintering occurs by diffusion of atoms through the microstructure. This diffusion is caused by a gradient of chemical potential – atoms move from an area of higher chemical potential to an area of lower chemical potential. The different paths the atoms take to get from one spot to another are the sintering mechanisms. The six common mechanisms are:

- Surface diffusion – Diffusion of atoms along the surface of a particle

- Vapor transport – Evaporation of atoms which condense on a different surface

- Lattice diffusion from surface – atoms from surface diffuse through lattice

- Lattice diffusion from grain boundary – atom from grain boundary diffuses through lattice

- Grain boundary diffusion – atoms diffuse along grain boundary

- Plastic deformation – dislocation motion causes flow of matter

Also one must distinguish between densifying and non-densifying mechanisms. 1–3 above are non-densifying – they take atoms from the surface and rearrange them onto another surface or part of the same surface. These mechanisms simply rearrange matter inside of porosity and do not cause pores to shrink. Mechanisms 4–6 are densifying mechanisms – atoms are moved from the bulk to the surface of pores thereby eliminating porosity and increasing the density of the sample.

Grain Growth

A grain boundary(GB) is the transition area or interface between adjacent crystallites (or grains) of the same chemical and lattice composition. The adjacent grains do not have the same orientation of the lattice thus giving the atoms in GB shifted positions relative to the lattice in the crystals. Due to the shifted positioning of the atoms in the GB they have a higher energy state when compared with the atoms in the crystal lattice of the grains. It is this imperfection that makes it possible to selectively etch the GBs when one wants the microstructure visible. Striving to minimize its energy leads to the coarsening of the microstructure to reach a metastable state within the specimen. This involves minimizing its GB area and changing its topological structure to minimize its energy. This grain growth can either be normal or abnormal, a normal grain growth is characterized by the uniform growth and size of all the grains in the specimen. Abnormal grain growth is when a few grains grow much larger than the remaining majority.

Grain Boundary Energy/Tension

The atoms in the GB are normally in a higher energy state than their equivalent in the bulk material. This is due to their more stretched bonds, which gives rise to a GB tension σ_{GB}. This extra energy that the atoms possess is called the grain boundary energy, γ_{GB}. The grain will want to minimize this extra energy thus striving to make the grain boundary area smaller and this change requires energy.

"Or, in other words, a force has to be applied, in the plane of the grain boundary and acting along a line in the grain-boundary area, in order to extend the grain-boundary area in the direction of the force. The force per unit length, i.e. tension/stress, along the line mentioned is σGB. On the basis of this reasoning it would follow:

$$\sigma_{GB}dA \text{ (work done)} = \gamma_{GB}dA \text{ (energy change)}$$

with dA as the increase of grain-boundary area per unit length along the line in the grain-boundary area considered."

The GB tension can also be thought of as the attractive forces between the atoms at the surface and the tension between these atoms is due to the fact that there is a larger interatomic distance between them at the surface compared to the bulk (i.e. surface tension). When the surface area becomes bigger the bonds stretch more and the GB tension increases. To counteract this increase in tension there must be a transport of atoms to the surface keeping the GB tension constant. This diffusion of atoms accounts for the constant surface tension in liquids. Then the argument,

$$\sigma_{GB}dA \text{ (work done)} = \gamma_{GB}dA \text{ (energy change)}$$

holds true. For solids, on the other hand, diffusion of atoms to the surface might not be sufficient and the surface tension can vary with an increase in surface area. For a solid, one can derive an expression for the change in Gibbs free energy, dG, upon the change of GB area, dA. dG is given by:

$$\sigma_{GB}dA \text{ (work done)} = dG \text{ (energy change)} = \gamma_{GB}dA + Ad\gamma_{GB}$$

which gives:

$$\sigma_{GB} = \gamma_{GB} + \frac{Ad\gamma_{GB}}{dA}$$

σ_{GB} is normally expressed in units of $\frac{N}{m}$ while γ_{GB} is normally expressed in units of $\frac{J}{m^2}$ $(J = Nm)$ since they are different physical properties.

Mechanical Equilibrium

In a two-dimensional isotropic material the grain boundary tension would be the same for the grains. This would give angle of 120° at GB junction where three grains meet. This would give the structure a hexagonal pattern which is the metastable state (or mechanical equilibrium) of the 2D specimen. A consequence of this is that to keep trying to be as close to the equilibrium as possible. Grains with fewer sides than six will bend the GB to try keep the 120° angle between each other. This results in a curved boundary with its curvature towards itself. A grain with six sides will, as mentioned, have straight boundaries while a grain with more than six sides will have curved boundaries with its curvature away from itself. A grain with six boundaries (i.e. hexagonal structure) are in a metastable state (i.e. local equilibrium) within the 2D structure. In three dimensions structural details are similar but much more complex and the metastable structure for a grain is a non-regular 14-sided polyhedra with doubly curved faces. In practice all arrays of grains are always unstable and thus always grows until its prevented by a counterforce.

Grains strive to minimize their energy, and a curved boundary has a higher energy than a straight boundary. This means that the grain boundary will migrate towards the curvature. The consequence of this is that grains with less than 6 sides will decrease in size while grains with more than 6 sides will increase in size.

Grain growth occurs due to motion of atoms across a grain boundary. Convex surfaces have a higher chemical potential than concave surfaces therefore grain boundaries will move toward their center of curvature. As smaller particles tend to have a higher radius of curvature and this results in smaller grains losing atoms to larger grains and shrinking. This is a process called Ostwald ripening. Large grains grow at the expense of small grains. Grain growth in a simple model is found to follow:

$$G^m = G_0^m + Kt$$

Here G is final average grain size, G_o is the initial average grain size, t is time, m is a factor between 2 and 4, and K is a factor given by:

$$K = K_0 e^{\frac{-Q}{RT}}$$

Here Q is the molar activation energy, R is the ideal gas constant, T is absolute temperature, and K_o is a material dependent factor.

Reducing Grain Growth

Solute Ions

If a dopant is added to the material (example: Nd in BaTiO$_3$) the impurity will tend to stick to the grain boundaries. As the grain boundary tries to move (as atoms jump from the convex to concave surface) the change in concentration of the dopant at the grain boundary will impose a drag on the boundary. The original concentration of solute around the grain boundary will be asymmetrical in most cases. As the grain boundary tries to move the concentration on the side opposite of motion will have a higher concentration and therefore have a higher chemical potential. This increased chemical potential will act as a backforce to the original chemical potential gradient that is the reason for grain boundary movement. This decrease in net chemical potential will decrease the grain boundary velocity and therefore grain growth.

Fine second phase particles

If particles of a second phase which are insoluble in the matrix phase are added to the powder in the form of a much finer powder than this will decrease grain boundary movement. When the grain boundary tries to move past the inclusion diffusion of atoms from one grain to the other will be hindered by the insoluble particle. Since it is beneficial for particles to reside in the grain boundaries and they exert a force in opposite direction compared to the grain boundary migration. This effect is called the Zener effect after the man who estimated this drag force to:

$$F = \pi r \lambda \sin(2\theta)$$

where r is the radius of the particle and λ the interfacial energy of the boundary if there are N particles per unit volume their volume fraction f is:

$$f = \frac{4}{3}\pi r^3 N$$

assuming they are randomly distributed. A boundary of unit area will intersect all particles within a volume of 2r which is 2Nr particles. So the number of particles n intersecting a unit area of grain boundary is:

$$n = \frac{3f}{2\pi r^2}$$

Now assuming that the grains only grow due to the influence of curvature, the driving force of growth is $\frac{2\lambda}{R}$ where (for homogeneous grain structure) R approximates to the mean diameter of the grains. With this the critical diameter that has to be reached before the grains ceases to grow:

$$nF_{max} = \frac{2\lambda}{D_{crit}}$$

This can be reduced to $D_{crit} = \frac{4r}{3f}$ so the critical diameter of the grains is dependent of the size and volume fraction of the particles at the grain boundaries.

It has also been shown that small bubbles or cavities can act as inclusion.

More complicated interactions which slow grain boundary motion include interactions of the surface energies of the two grains and the inclusion and are discussed in detail by C.S. Smith.

Natural Sintering in Geology

Petrifying spring in Réotier near Mont-Dauphin, France.

In geology a natural sintering occurs when a mineral spring brings about a deposition of chemical sediment or crust, for example as of porous silica.

A sinter is a mineral deposit that presents a porous or vesicular texture; its structure shows small cavities. These may be siliceous deposits or calcareous deposits.

Siliceous sinter is a deposit of opaline or amorphous silica which appears as incrustations near hot springs and geysers. It sometimes forms conical mounds, called geyser cones, but can also form as a terrace. The main agents responsible for the deposition of siliceous sinter are algae and other vegetation in the water. Altering of wall rocks can also form sinters near fumaroles and in the deeper channels of hot springs. Examples of siliceous sinter are geyserite and fiorite. They can be found in many places, including Iceland, El Tatio geothermal field in Chile, New Zealand, and Yellowstone National Park and Steamboat Springs in the USA.

Calcareous sinter is also called tufa, calcareous tufa, or calc-tufa. It is a deposit of calcium carbonate, as with travertine. Called petrifying springs, they are quite common in limestone districts. Their calcareous waters deposit a sintery incrustation on surrounding objects. The precipitation is assisted with mosses and other vegetable structures, thus leaving cavities in the calcareous sinter after they have decayed.

Petrifying spring at Pamukkale, Turkey.

Sintering of Catalysts

Sintering is an important cause for loss of catalyst activity, especially on supported metal catalysts. It decreases the surface area of the catalyst and changes the surface structure. For a porous catalytic surface, the pores may collapse due to sintering, resulting in loss of surface area. Sintering is in general an irreversible process.

Small catalyst particles (which have the highest relative surface areas) and a high reaction temperature are in general both factors that increase the reactivity of a catalyst. However, these factors are also the circumstances under which sintering is occurring. Specific materials may also increase the rate of sintering. On the other hand, by alloying catalysts with other materials, sintering can be reduced. Especially rare earth metals have been shown to reduce sintering of metal catalysts when alloyed.

For many supported metal catalysts, sintering starts to become a significant effect at temperatures over 500 °C (932 °F). Catalysts that operate at higher temperatures, such as a car catalyst, use structural improvements to reduce or prevent sintering. These improvements are in general in the form of a support made from an inert and thermally stable material such as silica, carbon or alumina.

References

- C. Barry Carter, M. Grant Norton (2007). Ceramic Materials: Science and Engineering. Springer Science+Business Media, LLC. pp. 427–443. ISBN 978-0-387-46270-7

- G. Kuczynski (6 December 2012). Sintering and Catalysis. Springer Science & Business Media. ISBN 978-1-4684-0934-5

- I. Chorkendorff; J. W. Niemantsverdriet (6 March 2006). Concepts of Modern Catalysis and Kinetics. John Wiley & Sons. ISBN 978-3-527-60564-4

- Tuan, W.H.; Guo, J.K. (2004). "Multiphased ceramic materials: processing and potential". Springer. ISBN 3-540-40516-X

- Kingery, W. David; Bowen, H. K.; Uhlmann, Donald R. (April 1976). "Introduction to Ceramics" (2nd ed.). John Wiley & Sons, Academic Press. ISBN 0-471-47860-1

- Greene, Eric S. (20 October 2006). "Mass transfer in graded microstructure solid oxide fuel cell electrodes". Journal of Power Sources - Volume 161, Issue 1. Elsevier B.V. pp. 225–231. Retrieved 10, May 2020

- Kang, Suk-Joong L. (2005). Sintering: Densification, Grain Growth, and Microstructure. Elsevier Ltd. pp. 9–18. ISBN 978-0-7506-6385-4

- Deville, Sylvain (March 2008). "Freeze-Casting of Porous Ceramics: A Review of Current Achievements and Issues". Advanced Engineering Materials - Vol 10 Issue 3. John Wiley & Sons, Inc. pp. 155–169. Retrieved 19, April 2020

- Deville, Sylvain (April 2007). "Ice-templated porous alumina structures". Acta Materialia - Volume 55, Issue 6. Elsevier B.V. pp. 1965–1974. Retrieved 09, May 2020

Permissions

Index

CPSIA information can be obtained
at www.ICGtesting.com
Printed in the USA
BVHW062001260822
645617BV00004B/167

9 781639 891009